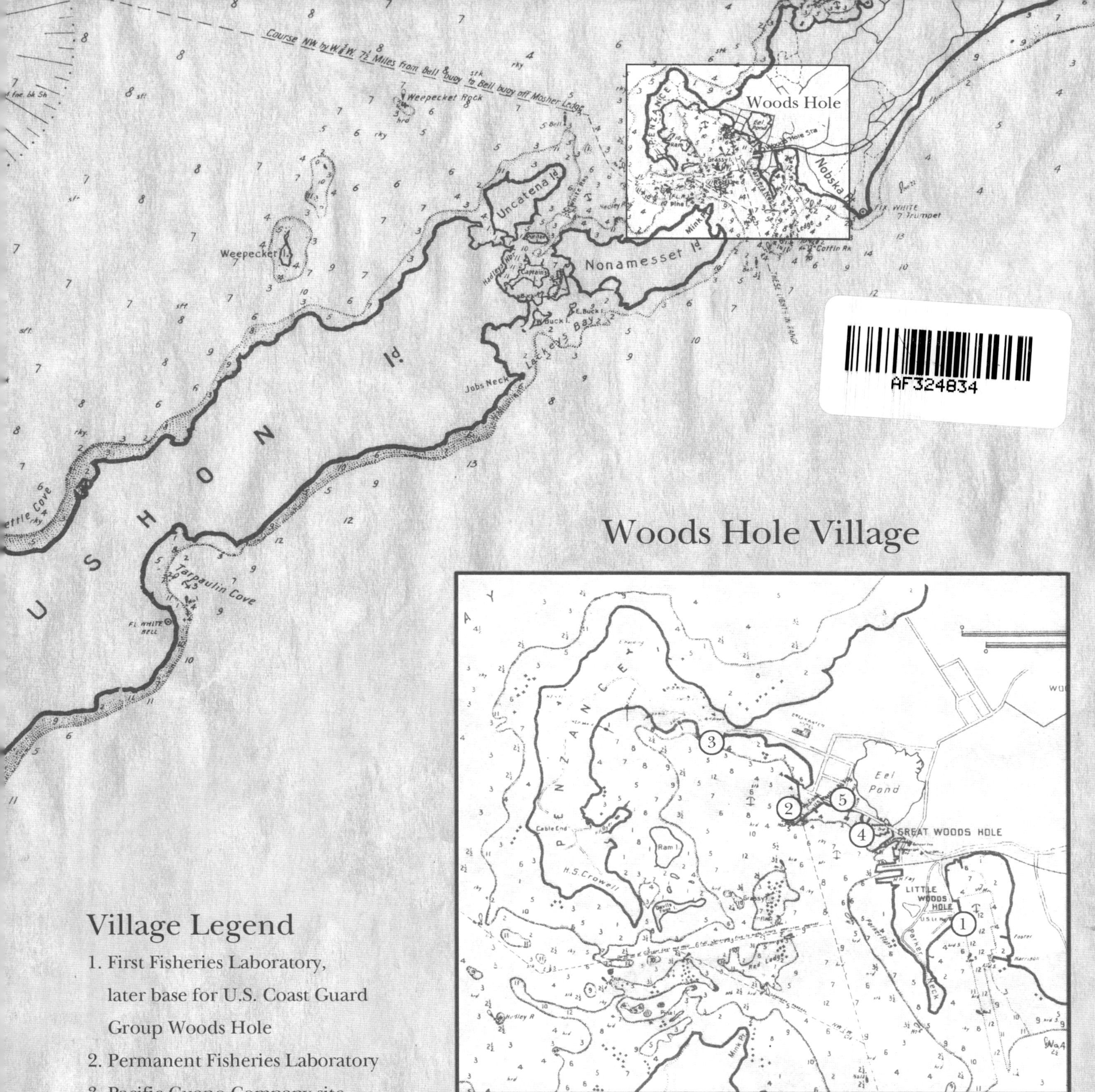

Woods Hole Village

Village Legend

1. First Fisheries Laboratory, later base for U.S. Coast Guard Group Woods Hole
2. Permanent Fisheries Laboratory
3. Pacific Guano Company site
4. Bar Neck Wharf (whaling dock), later marine facilities of the Woods Hole Oceanographic Institution
5. Marine Biological Laboratory

Anne-Marie
Best Wishes &
Happy Trails
MK

Anne-Marie—
All the best!
Debbie

Anne-Marie,
I know you'll love the
warm weather. Good luck
to you. Maureen

AM
Best of everything to
you — in your "new, warm"
life in the South! Keep
in touch — and fly back for
the lunch at the "Cookie"!
Blessings!
Maryanne

Dear Anne-Marie,
I hope you think of us when
you see this book — while having
a cocktail + enjoying the Southern
warmth! We'll miss you.
Stay in touch
Love,
Hazel

Anne-Marie!
Good luck.
keep in touch.
Anne

Anne-Marie
Best wishes
on your new
venture. Penny

Anne-Marie!
It was so nice
to meet you and
are you @ our
Christmas celebration!
Stay well and happy!
Shirley

Most human history has not afforded men much chance to pursue their curiosity, except as a hobby of the rich or within the refuge of a monastery. We can count ourselves fortunate to live in a society and at a time when we are actually paid to explore the universe.

—Henry Stommel, upon receiving the Bigelow Medal, 1974

Down to the Sea for Science

75 Years of Ocean Research, Education, and Exploration
at the Woods Hole Oceanographic Institution

VICKY CULLEN

When one picks up a fish, one may be said, allegorically, to hold one of the knots in an endless web of netting of which the countless other knots represent other facts, whether of marine chemistry, physics or geology, or other animals or plants. And just as one cannot make a fish-net until one has tied all the knots in their proper positions, so one cannot hope to comprehend this web until one can see its internodes in their true relationship.

 —Henry Bryant Bigelow, "A Developing Viewpoint in Oceanography," *Science,* January 24, 1930

Copyright © 2005 Woods Hole Oceanographic Institution

Woods Hole Oceanographic Institution
Fenno House, MS #40
Woods Hole, Massachusetts 02543

Edited and produced by Vern Associates, Inc.
Designed by Benjamin A. Jenness for Vern Associates, Inc.
Typeset by dix!
Printed and bound in China by C & C Offset Printing Co., Ltd.

Quotations on pages 84–85 and 114 from *Agassiz's Legacy: Scientists' Reflections on the Value of Field Experience*, by Elizabeth Higgins Gladfelter, copyright © 2002 by Oxford University Press, Inc., are used by permission of Oxford University Press, Inc.

Quotations on pages 8, 11, 82, and 94 from *Oceanography: The Past*, edited by M. Sears and D. Merriman, copyright © 1980 by Springer-Verlag New York, Inc., are used by permission of Springer-Verlag, Inc. Article titles and authors are cited in the endnotes.

Quotations on pages 19, 24, 39, 70, and 93 from *Biographical Memoirs*, copyright © National Academy of Sciences, are used by permission of the National Academies Press, Washington, D. C. These contributions are cited in the endnotes.

Quotations on pages 29, 30, and 38–39 from *On Almost Any Wind*, by Susan Schlee, copyright © 1978 by Cornell University, are used by permission of Cornell University Press.

Quotations from letters and notes held in the collections of the Rockefeller Archive Center and the Harvard University Archives and fully cited in the endnotes are used by permission of these institutions.

Photo and illustration credits:
Most of the photographs in *Down to the Sea for Science* were drawn from the Woods Hole Oceanographic Institution Archives collection. If no credit is listed below, the photo or illustration is from this archive. Credits for other sources and for known photographers of WHOI Archives photos follow, with page numbers in bold type:
i painting by Barbara Whitehead; **ii** Don Fay; **vi** Susan Tarbell; **2** (top) Northeast Fisheries Science Center (NFSC) Archives, (bottom) Baldwin Coolidge, from a postcard in the Marine Biological Laboratory (MBL) Archives Collection; **3** etching from an 1877 U. S. Commission of Fisheries report, courtesy Woods Hole Historical Collection (WHCC); **4** (top) NFSC Archives, (bottom) from HMS *Challenger* Report, Vol. 1, courtesy MBL Archives; **5** Baldwin Coolidge, courtesy MBL Archives; **6** (top) Baldwin Coolidge, (bottom) Henry Bigelow, both courtesy MBL Archives; **7** (bottom), NFSC Archives; **9** and **10** (top) Baldwin Coolidge, courtesy MBL Archives; **10** (bottom) courtesy of the Rockefeller Archive Center; **13** courtesy of United States Naval Institute Photo Archives, Annapolis; **16–17** W.C. Cullen; **18** (photo) John Welsh, (clipping) Center for American History, University of Texas at Austin; **20** Charles M. Weiss; **24** John Welsh; **28** (top) George Clarke, (middle) Daniel Merriman, (bottom) John Welsh; **29** (bottom) John Welsh; **32** Isaac Harter, Jr.; **33, 36,** and **37** Scott Bray; **45** courtesy WHCC; **46** Canadian Naval Research Establishment; **49** Don Fay; **50–51** Jan Hahn; **57** David Owen; **65** (bottom) Claude Ronne; **66** and **73** (top) Mary Curtis Cobb Thayer; **74** (top) David Owen, (bottom) Don Fay; **75** (top) Claude Ronne; **80** John Welsh; **81** Don Fay; **87** Redwood Wright; **92** (bottom) Clark M. Weiss; **93** (top) Time Life Pictures/Getty Images; **97** (bottom) Vicky Cullen; **99** (map) Jayne Doucette–WHOI Graphics, (cartoon) Conrad Neumann; **101** Bill Lambert; **108** Vicky Cullen; **109** (bottom) Danielle Fino; **113** (top) courtesy of Richard Backus, (bottom) Beecher Wooding; **114** courtesy of Carl Bowin; **117** (bottom) Fritz Goro, *Life,* courtesy of WHOI Archives; **119** (bottom left) Jan Hahn, (bottom right) Peter Wiebe; **120** (top) Deep Sea Drilling Program, (bottom) David Owen; **124** (bottom) *Bay City Times;* **127** Tom Kleindinst; **128** (top) Jan Hahn, (middle and bottom) Dean Conger © National Geographic Society; **130** (top) Vicky Cullen, (bottom) Frank Medeiros; **131** (top) Michael Schofield; **133** (bottom) courtesy of George Tupper; **135** (top) Frank Medeiros, (bottom) Vicky Cullen; **137** Jack Donnelly; **138** Vicky Cullen; **139** Shelley Dawicki; **140** WHOI Buoy Project; **141** (top) Mel Briscoe, (bottom) Upper Ocean Processes Group; **142** (top) Larry Madin, (bottom) Henry Bigelow, courtesy MBL Archives; **143** Larry Madin; **144–145** Christopher Knight; **146** Richard Bowen; **147** (top) Shelley Dawicki, (bottom) Keith von der Heydt; **148** (top) Shelley Dawicki, (bottom) Dave Gray; **149** Larry Workman; **150** (bottom) Doug Weisman; **151** Jayne Doucette–WHOI Graphics; **152** (top) Shelley Dawicki, (bottom) Dan Fornari; **153** © Crown Copyright 1998; **154** (top left) Dan Fornari, (top right) Jack Cook–WHOI Graphics, (bottom) Tom Kleindinst; **155** (top) Jayne Doucette, (bottom) Tom Kleindinst; **156** Walker O. Cain; **157** (top) Shelley Dawicki, (bottom) Terri Corbett; **158** (top) A Bird's Eye View, (bottom) Tom Kleindinst; **159** (top) Vicky Cullen, (bottom) Tom Kleindinst; **161** Amy Rader

Contents

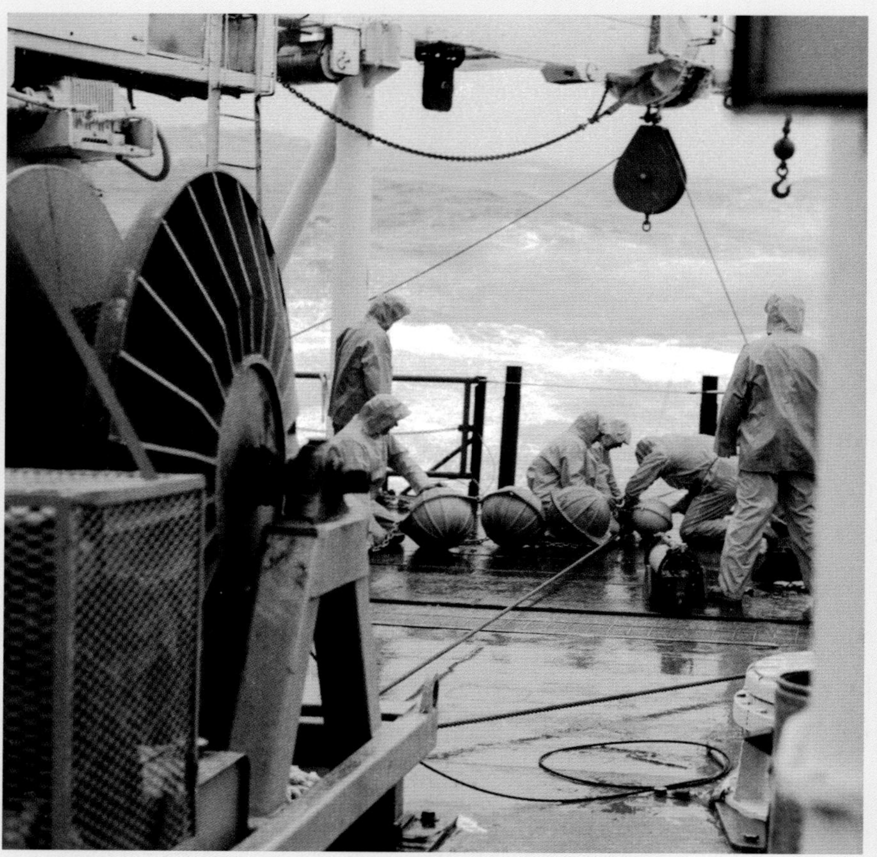

Members of the buoy group assemble a mooring aboard *Knorr* on a raw day in 1975.

They that go down to the sea in ships, that do business in great waters;
These see the works of the Lord, and his wonders in the deep.

—Psalm 107:23–24

Introduction

I am proud to be the seventh director of the Woods Hole Oceanographic Institution and eager to support its continuation as a world leader in ocean science. Reading over this history, I am struck by WHOI's steady growth, its underlying theme of excellence at sea, and the personal and scientific threads that both bind us to our ancestors and provide a springboard to the future. As a thirty-three-year WHOI veteran myself, I have experienced the pains and pleasures of science at sea, benefited from the enormous talents of our excellent support staff, and collaborated with gifted oceanographers, engineers, and students—and been privileged to know many of the Institution's colorful characters.

As we lay the groundwork for the next seventy-five years, our vision is to sustain leadership in oceanographic research and higher education by nurturing inventive minds in a creative environment. We do this by providing unmatched access to the sea and premier shore-based facilities and by conveying the benefits of scientific discovery to a broad audience eager for knowledge of the oceans.

Founding WHOI director Henry Bigelow consulted scientists near and far to lay the early-twentieth-century foundations for the growth of U.S. oceanography and a new institution on the East Coast. Similarly, I work with colleagues at other marine science laboratories around the country toward a common goal: to create a rising tide of support for U.S. oceanography and reveal untold new discoveries to benefit all the world.

Robert B. Gagosian
WHOI President and Director
2005

The logo at left was designed for the Woods Hole Oceanographic Institution in 1961 by Gale Pasley and Mary Minot of the Institution's graphics group. By 1970, when it was first printed on the cover of the annual report, the logo had been streamlined to the form at right, generally used for the next 35 years. A hybrid embossed logo appeared on the cover of the 1968 annual report, and, for digitization purposes, a more angular form is sometimes used.

Acknowledgments

Thanks are due most of all to WHOI President and Director Bob Gagosian for supporting this history of the Woods Hole Oceanographic Institution, for reviewing the manuscript, and for allowing a candid historical discussion. Historian Paul Lucier made copious notes on the 1960s, 1970s, and 1980s; they have served as a fine source of information. Mary Schumacher did the research for the 1990s, carefully reviewed the manuscript, and made excellent suggestions.

Dick Backus, Dick Edwards, and Charlie Innis were founts of information and most helpful reviewers. Other manuscript and page proof reviewers included Shelley Dawicki, Bob Ducommun, John Farrington, Jim Kent, Jim Luyten, Jane Neumann, Jake Peirson, Jeannine Pires, Karen Rauss, Audrey Rogerson, Dave Ross, Tony Ryan, Dan Stuermer, Mildred Teal, Freddy Valois, and Hoyt Watson.

It is fortunate that, over the years, designated archivists—Norman Allen, Stuart Culy, Bill Dunkle, Margot Garritt, Marisa Hudspeth, and Stan Reeves—and their colleagues have managed to maintain a rich archive. In addition to official documents, letters, and photos, the WHOI Archives contain, among many other things, a historic instrument collection and more than seventy-five oral histories, most of them recorded on a volunteer basis by former science teacher Frank Taylor, who has driven thousands of miles between his home in Framingham and Woods Hole to conduct these interviews (not to mention a visit to Hawaii to interview Al Woodcock, an insightful member of the original *Atlantis* crew and later a scientist). I am grateful for the cheerful assistance with text and photo research rendered by the WHOI Data Library and Archives staff—Marisa Hudspeth, Robin Hurst, Colleen Hurter, Ellen Levy, and Lisa Raymond—and also to Susan Witzell at the Woods Hole Historical Collection and to Jennifer Walton at the MBL/WHOI Library. Dave Gray of WHOI Graphic Services supplied the excellent digital images that grace these pages.

Brian Hotchkiss and Peter Blaiwas of Vern Associates, book packagers, have been a joy to work with as the book came together, lending their advice and experience to shape *Down to the Sea for Science* into a creditable publication.

There have been, of course, many others who allowed themselves to be interviewed or queried, or who offered interest and encouragement along the way. Finally, thanks and love to my husband, John Waterbury, and sons, Andrew and Matthew, for putting up with more than a year's worth of a very distracted wife and mother.

Vicky Cullen
Woods Hole, Massachusetts
April 2005

Preface

Sit down to lunch any day at the Woods Hole Oceanographic Institution's Buttery, and you are likely to be treated to a few amusing (or hair-raising) sea stories. The tie that binds this institution is its seagoing tradition, born in the summer of 1931 with the maiden voyage of WHOI's first research vessel, *Atlantis,* crossing from Copenhagen to Boston. If the storytellers, usually shy about the idea, could be persuaded to write down their tales, they would fill volumes. A few of their stories are captured here along with, we hope, a sense of an institution designed to nurture scientific creativity, determination, and innovation.

By the eleventh of its seventy-five years, WHOI was no longer a small laboratory. The story of its scientific achievements—and they are legion—is told in the scientific literature; this small volume highlights a few of them. Six science story lines that can be traced more or less from the beginning of the Institution are offered as features to represent the many others there wasn't room to include. Likewise, the names of only a few people who have contributed significantly to the still-maturing science of oceanography would fit into the limits of this book, which focuses largely on the early, formative years of the Institution. The narrative

Work at sea is not easy (*Atlantis*, 1950s).

depicting the later years of an increasingly complex Institution is more fragmentary—left for fuller treatment in a future volume.

The chronological narrative begins not with the founding of the Institution in 1930, but with significant earlier events and circumstances that influenced its character and development.

A 1901 expedition to the Maldive Islands (map) and an intensive 1912–1916 investigation of the Gulf of Maine (photo) formed Henry Bigelow as an oceanographer.

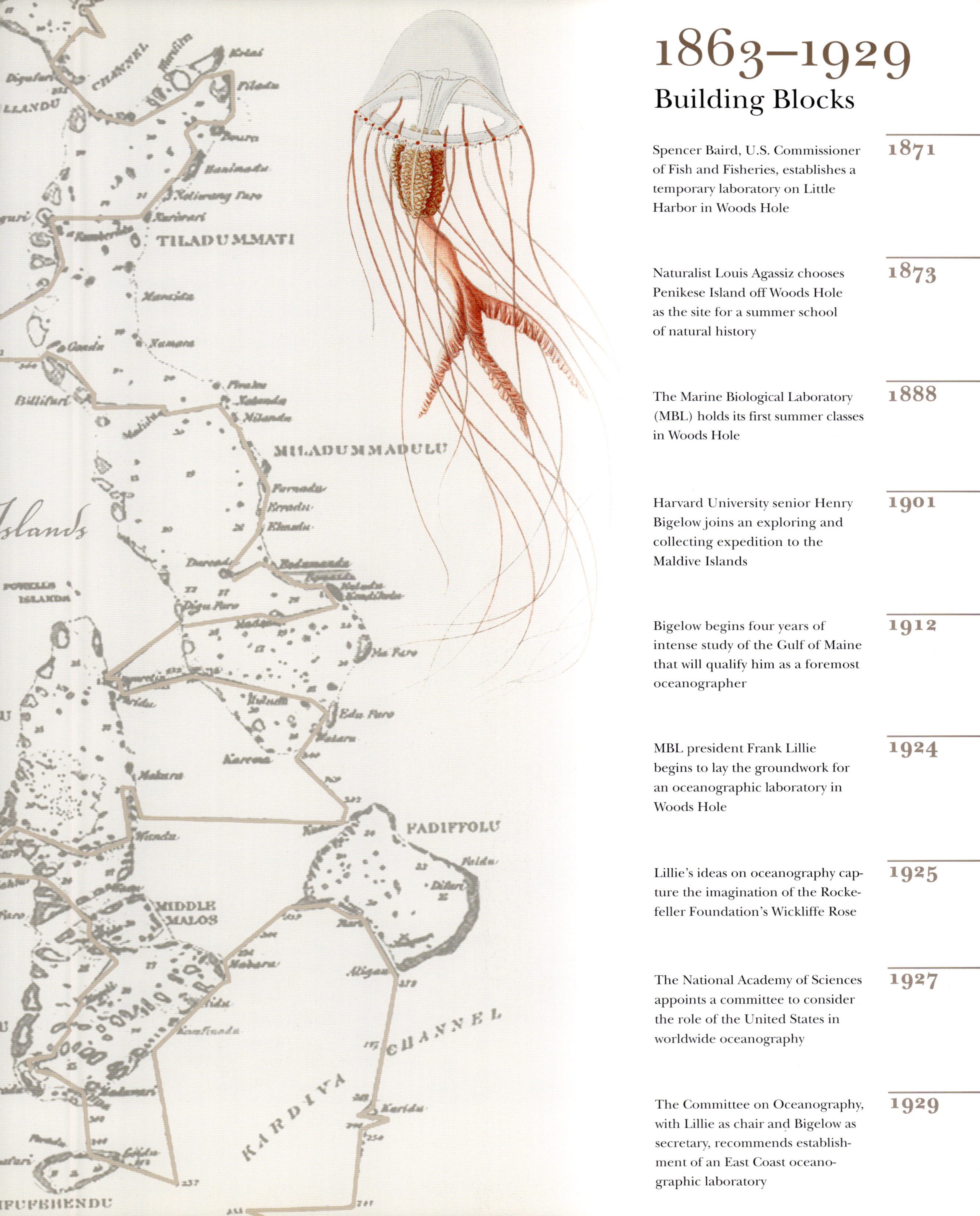

1863–1929
Building Blocks

1871 — Spencer Baird, U.S. Commissioner of Fish and Fisheries, establishes a temporary laboratory on Little Harbor in Woods Hole

1873 — Naturalist Louis Agassiz chooses Penikese Island off Woods Hole as the site for a summer school of natural history

1888 — The Marine Biological Laboratory (MBL) holds its first summer classes in Woods Hole

1901 — Harvard University senior Henry Bigelow joins an exploring and collecting expedition to the Maldive Islands

1912 — Bigelow begins four years of intense study of the Gulf of Maine that will qualify him as a foremost oceanographer

1924 — MBL president Frank Lillie begins to lay the groundwork for an oceanographic laboratory in Woods Hole

1925 — Lillie's ideas on oceanography capture the imagination of the Rockefeller Foundation's Wickliffe Rose

1927 — The National Academy of Sciences appoints a committee to consider the role of the United States in worldwide oceanography

1929 — The Committee on Oceanography, with Lillie as chair and Bigelow as secretary, recommends establishment of an East Coast oceanographic laboratory

The first marine biological station in the world was established in 1842 at Ostend, Belgium.[1] By 1871, when Spencer Baird created his first, temporary fisheries laboratory in Woods Hole, there was broad interest in seaside studies, and a number of marine stations had been established. The most international of these, the Zoological Station of Naples, Italy, was founded in 1872. Marine stations elsewhere in the United States included Stanford University's Hopkins Seaside Laboratory, which held its first classes in 1892; the Marine Biological Institution of San Diego (now the Scripps Institution of Oceanography at the University of California, San Diego), formally established in 1903; and the Puget Sound Biological Station (now the University of Washington's Friday Harbor Laboratories), which began operations in 1910. The focus of these stations was coastal. When the Woods Hole Oceanographic Institution was founded in 1930, no open-ocean research vessels were available to American scientists.

Spencer Baird, the first U.S. commissioner of fish and fisheries, set up a temporary laboratory in Woods Hole in 1871 and established a permanent fisheries station in the village in 1885.

The Candle House, at right in this photo from about 1890, is all that remains of mid-nineteenth-century whaling in Woods Hole. The smaller building at center housed the try works, where whale oil was rendered. Today the Candle House is used by the Marine Biological Laboratory (MBL) for offices. The view is from the present location of MBL's Crane Building.

The Pacific Guano Company manufactured fertilizer on Penzance Point in the late nineteenth century.

The earliest historical records for Woods Hole, Massachusetts, concern purchase of land from the Wampanoag Indian Job Notantico by fourteen settlers on December 30, 1679. Small-scale farming, some fishing, and, later, saltworks were the first industries. By 1790, the village boasted ten houses and a population of seventy-four. Sixty years later, there were thirty-four houses and two hundred people. Whaling ships (peaking at nine) worked out of Woods Hole from 1828 to 1864, and the Pacific Guano Company produced fertilizer (based on local fish and bird droppings collected on islands in the Pacific Ocean) from 1863 to 1889 at its factory on the peninsula then known as "Long Neck," now Penzance Point. Four years before Pacific Guano ceased operations, Baird's location of a permanent fisheries laboratory in Woods Hole initiated the scientific bent of this delightful village.

There are many marine laboratories in the world, but there is only one Woods Hole. . . . The geographical setting; the remoteness from large centers of population, which insures freedom from pollution and maintenance of purity of the waters; the variety of shore conditions and animal habitats; and ready access to the open ocean—these explain why successively the United States Bureau of Fisheries (1885), the Marine Biological Laboratory (1888), and the Woods Hole Oceanographic Institution (1930) selected Woods Hole as the site for their scientific laboratories.

—Frank Rattray Lillie, The Woods Hole Marine Biological Laboratory, 1944

Spencer Baird, left, and colleagues push off for an early Woods Hole collecting trip.

The 1872–1876 circumnavigation by the 200-foot (61-meter), sail-and-steam-powered corvette HMS *Challenger* resulted in vast sample collections. The observations of its biologists, chemists, and physicists filled fifty volumes.

1863 Fish zoologist and assistant secretary of the Smithsonian Institution Spencer Fullerton Baird makes his first visit to Woods Hole. Later, in 1871, as newly named U.S. Commissioner of Fish and Fisheries, Baird returns to Woods Hole to establish a temporary laboratory on Little Harbor as a base for collecting fish. After working in a number of other coastal areas, Baird chooses Woods Hole as a permanent location, solicits contributions from individuals and institutions for purchase of land, presents the land to the government, and, in 1885, constructs the buildings of the Woods Hole station of the Bureau of Fisheries.

1872 On December 21, HMS *Challenger* departs Plymouth, England, beginning a three-and-a-half-year circumnavigation of the world that is often said to mark the birth of modern oceanography. It is one of several similar mid-nineteenth-century "grand surveying expeditions" that collect thousands of animals and seafloor deposits. But each cruise is a "one off," conducted aboard chartered or navy ships from various countries for limited time periods, with no plans for further work on the open ocean after the samples are processed.

1873 Swiss naturalist Louis Agassiz, founding director of the Harvard University Museum of Comparative Zoology, establishes the Anderson School of Natural History on Penikese Island, one of the Elizabeth Islands off Woods Hole. Twenty-six men and seventeen women from six East Coast and five Midwestern states attend. The school, named for New York merchant John Anderson, who provided both the island and funding for the school, offers lectures, collecting trips, and laboratory space from July 8 to August 29. Agassiz's philosophy, posted on laboratory walls, admonishes students to "Study nature not books." The school operates for only two summers but leaves a lasting impression on its participants, and its spirit still lives in today's Woods Hole laboratories.

1881–1888

Inspired by the Anderson School, two Boston societies, the Women's Education Association and the Boston Society of Natural History, sponsor a summer program from 1881 to 1886 "to afford opportunities for the study of the development, anatomy, and habits of common types of marine animals" at a rented laboratory in Annisquam, Massachusetts. Pleased with its success, they transfer their interests and assets (glassware, boats, furniture, and fixtures) to a group charged with establishing a permanent seaside laboratory. This group incorporates as the Marine Biological Laboratory (MBL) in March 1888, chooses Woods Hole as the site, purchases a small plot of land near the Bureau of Fisheries laboratory, and constructs a small frame-shingled building that opens on July 17, 1888. Charles O. Whitman, an Anderson School student during both its sessions and now on the faculty of the newly founded Clark University in Worcester, Massachusetts, is the first director. His successor,

Though Louis Agassiz's Anderson School on Penikese Island lasted for only two years (the summers of 1873 and 1874), it inspired the development of later scientific establishments in Woods Hole. All the buildings pictured burned in 1892.

"Old Main" was one of the Marine Biological Laboratory's first buildings. It was constructed in three sections in 1888, 1890, and 1892.

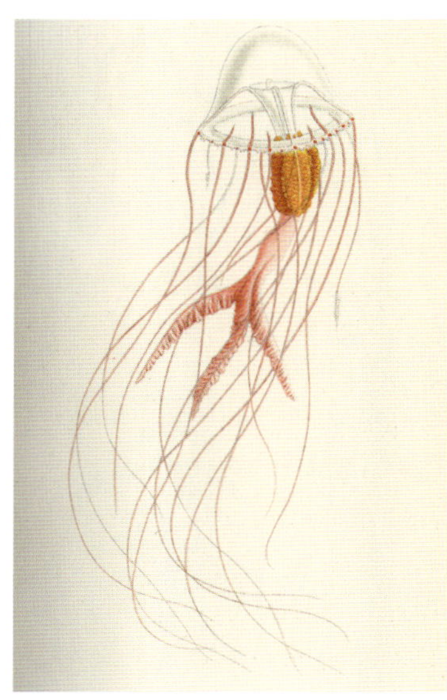

Drawing from Henry Bigelow's paper "Medusae from the Maldive Islands." Bigelow was in charge of collecting and curating gelatinous animals during Alexander Agassiz's 1901 voyage to the Maldives, an expedition conducted along the lines of the grand surveying cruises common in the nineteenth century.

Frank R. Lillie, later writes of him: "He had come under the influence of three of the greatest zoölogists of the time, Agassiz, Leuckart, and Dohrn;* was intimately acquainted with the work of the two most important marine laboratories, Penikese and Naples; and at forty-five years of age was full of energy and directed enthusiasm. He was also to prove a wise and inspiring leader."[2]

1901

Henry Bryant Bigelow, a senior at Harvard University studying invertebrate zoology, screws up the courage to ask Alexander Agassiz (Louis's son), current director of the Museum of Comparative Zoology, if he might join his exploring and collecting expedition to the Maldive Islands in the Indian Ocean. "[W]hile I hadn't the least idea where the Maldive Islands were," Bigelow writes later, "I decided I'd like to go along too!"[3] A positive reply from Agassiz initiates one of the most influential careers in American oceanography.[4]

1911

Bigelow completes a Ph.D. at Harvard in 1906 and by 1911 has made other cruises with Agassiz and worked on resulting samples. But now he is drifting, at "loose ends"[5] following Agassiz's death in 1910. Sir John Murray, a prominent Scottish oceanographer who sailed with *Challenger,* visits the United States to eulogize Agassiz and to encourage the U.S. government to support oceanographic research in the western North Atlantic.[6] He galvanizes Bigelow with the question "How much is known about the Gulf of Maine?"

1912–1916

Bigelow undertakes extensive studies of the Gulf of Maine on the Bureau of Fisheries schooner *Grampus.* By conducting

* Rudolph Leuckart was a highly regarded late-nineteenth-century professor at the University of Leipzig, and Anton Dohrn was the founder and first director of the Naples Zoological Station.

year-round physical, chemical, biological, and geological studies in a single body of water, he emulates the approach encouraged in Europe by the International Council for the Exploration of the Seas, organized in 1899 by the king of Denmark to investigate conditions in the North Atlantic "for the especial benefit of fisheries."[7] This work establishes Bigelow as a foremost oceanographer.

The National Academy of Sciences forms the National Research Council (NRC) to organize U.S. scientific research.[8]

1918

Norwegian explorer and oceanographer Fridtjof Nansen visits Washington, D.C., to advocate a stronger American effort in oceanography.[9]

1919

The National Academy of Sciences appoints Bigelow one of three members of its first Committee on Oceanography. However, neither broad interest nor financial support has yet been generated for oceanography; the committee accomplishes little and dissolves in 1923.

Henry Bigelow steers the U.S. Bureau of Fisheries vessel *Grampus* (*left*), about 1914. Named for a large dolphin, *Grampus* was built in 1886 to collect live fish and to exemplify an improved fishing schooner. While using the boat for surveys in the Gulf of Maine, Bigelow found that it was not sufficiently stable for oceanographic work, and its small hoisting engine could not handle heavy equipment. He applied his experience with *Grampus* to the design of WHOI's first research vessel, *Atlantis*.

The American Society of Zoologists selects Frank Lillie as representative to the NRC Division of Biology and Agriculture; by 1922, he will become division chair.

1923

William Ritter, founding director of the research station that is by this time called the Scripps Institution for Biological Research at the University of California, is preparing to retire, and the university offers Henry Bigelow the directorship. With career and family established on the East

Coast, he declines. T. Wayland Vaughan of the U.S. Geological Survey in Washington, D.C., a sedimentologist with strong interests in corals and foraminifera, takes the job.

1924 Frank Lillie begins to discuss "the share of the United States of America in a world wide program of Oceanographic Research"[10] with other North American marine scientists. In 1925, he takes up the matter with Wickliffe Rose, president of both the Rockefeller Foundation's domestic General Education Board and its International Education Board. Chair of the Department of Zoology at the University of Chicago and president of the Marine Biological Laboratory in Woods Hole, Lillie is a major voice in early-twentieth-century marine science. Rose commands the means to stimulate U.S. progress in oceanography. Lillie and Rose carry their discussions to the president of the Carnegie Institution and the chair of the National Research Council. This group determines that the "best method of progress" will be appointment of a committee on oceanography by the National Academy of Sciences "to consider the entire subject broadly."[11]

The U.S. Navy is also interested in oceanography. Following strong suggestions both from within the Navy and from civilian scientists that oceanographic investigations promise political, economic, and scientific rewards for the United States, the Navy convenes the Interagency Conference on Oceanography. Its mission is to suggest the most profitable application of naval and national resources for oceanographic exploration. Henry Bigelow represents the Department of Commerce on this committee, which meets with limited success because the Coolidge administration is not interested in oceanography.

1925 Bigelow and Willis Rich, director of the Bureau of Fisheries, visit Wickliffe Rose at the Rockefeller Foundation to encourage support for fisheries research. Rose asks them to prepare a ten-year oceanographic plan. Rose then travels six thousand miles during the month of July, visiting research stations all over the United States and Canada. In the fall, Bigelow sends Rose a plan "on work in Oceanography which can be accomplished by a suitably equipped ship."[12]

The Lillie and the Rose

Frank R. Lillie was the true "founding father" of the Woods Hole Oceanographic Institution. In the early 1920s, he thought it was time for the United States to play a larger role in international marine science, and he set about accomplishing this goal. Born in Toronto in 1870, Lillie did graduate work with Charles O. Whitman, first at Clark University and then at the newly established University of Chicago. By 1910, he succeeded Whitman as chair of the Chicago Department of Zoology and also as president of the Marine Biological Laboratory (MBL) in Woods Hole, Massachusetts. He was elected to the National Academy of Sciences (NAS) in 1915 and was the academy's president from 1935 to 1939, the first year serving simultaneously as chair of the National Academy of Engineering.

Lillie knew that planning committees needed support and that adequate funding would be essential to the success of a new oceanographic institution. He was well acquainted with the Rockefeller Foundation, having arranged its support for both his affiliate institutions, and he interested Wickliffe Rose, president of the foundation's General Education Board, in oceanography. John D. Rockefeller established the board

In 1893, Frank Lillie (far left) was a student in the embryology class taught by MBL director Charles Whitman (the white-haired gentleman at center). Lillie later succeeded Whitman as MBL Director and had a profound influence on both that institution and its oceanographic neighbor.

Frank R. Lillie was the driving force for the East Coast laboratory that became the Woods Hole Oceanographic Institution.

in 1903 to aid education in the United States "without distinction of race, sex, or creed." Rose contributed significantly to creating an educational system in his native southern United States, then turned his attention to health, notably the control of hookworm disease and yellow fever, and finally to "a search for new truths. What, he wanted to know, could be learned from the skies through the Palomar telescope, or from the seas in marine biological research institutes." Rose believed "the only way that mankind will find peace and contentment will be for men and women to determine and understand the nature of life."[13] Though he retired in 1928, Rose arranged for the foundation to fund both the NAS Committee on Oceanography and the new oceanographic institution it would recommend.

Lillie appears to have decided early in the discussions that Woods Hole was the right place for a new oceanographic laboratory designed to focus on the Atlantic Ocean. In the Rockefeller Foundation Archive Center notes regarding a luncheon with Lillie at Chicago's Hotel del Prado on October 5, 1927, less than six months after the Committee on Oceanography was appointed, Rose wrote, "In so far as one can visualize the results . . . in advance, it would seem that the study might lead to the Academy's undertaking the establishment of an Institute of Oceanography at Woods Hole."[14] Lillie thought this new laboratory should be patterned after MBL, hosting university professors and students each summer. Not everyone agreed on the location. Harvard professor G. H. Parker commented, "I think it would be difficult to keep men in Woods Hole for the winter."[15]

Wickliffe Rose was Lillie's principal early contact at the Rockefeller Foundation.

1927

With support from colleagues including Carnegie Corporation president John C. Merriam and Scripps director Vaughan, Lillie proposes a committee on oceanography at the National Academy of Sciences meeting in Washington, D.C., on April 27. "I must confess," he writes to Rose, "that I was surprised to find with what enthusiasm it was received." [16] In addition to Lillie (chair), Merriam, and Vaughan, the committee includes William Bowie, chief, Division of Geodesy, U.S. Coast and Geodetic Survey; E. G. Conklin, professor of zoology, Princeton University, and president of the Bermuda Biological Station for Research, Inc.; and B. M. Duggar, professor of plant physiology, University of Wisconsin, and head of MBL's Department of Botany.

Early in 1927 . . . Frank Lillie found the consensus he needed to organize oceanography. More than a ship was necessary, for a ship required a formal program, which in turn required an organizational structure to carry it out. The Marine Biological Laboratory did not have such a structure. Rather it accommodated individuals who brought with them to Woods Hole their own independent research problems. For shipboard research, scientists would have to sacrifice their customary individualism in favor of concerted scheduling. Though their problems might remain individual, they would have to be integrated into a program that took the ship from one part of the ocean to another. Moreover, so little was known about the North Atlantic that the ship would for the first few years spend much of its time in routinely gathering descriptive data. Such a program could not interest the scientists who came to the Marine Biological Laboratory. If Lillie's ambition was to base the new vessel in Woods Hole, some institution other than the Marine Biological Laboratory or the Bureau of Fisheries would have to operate it.[17]

—Harold Burstyn, *Oceanography: The Past,* 1980

The Bureau of Fisheries publishes the last of Henry Bigelow's three book-length monographs on Gulf of Maine fishes (1925), plankton (1926), and physical oceanography (1927). *Fishes of the Gulf of Maine* is coauthored by W. W. Welsh, who died during preparation of the monograph; Bigelow is sole author of the other two. Together these works render the Gulf of Maine possibly the best-known body of water of comparable size in the world—certainly the region most thoroughly explored by individual effort. British fisheries biologist Michael Graham later writes, "For one man to have made such a clear and complete job of a relatively large area . . . was a monumental job of which any man could be proud even if he had done nothing else in his whole life." [19]

Flounder drawing from Bigelow's early work, *Fishes of the Gulf of Maine* [18]

1928

The Rockefeller Foundation appropriates $75,000 to cover expenses of the Committee on Oceanography and travel for committee advisers. The latter include Norwegian oceanographer Harald U. Sverdrup and A. G. Huntsman, University of Toronto professor and director of the St. Andrews, New Brunswick, Marine Biological Station. Summer meetings in Woods Hole benefit from the wisdom of Bureau of Fisheries and MBL visiting scientists, whose home bases range from the California Institute of Technology to Yale and Columbia universities.

The Committee on Oceanography invites Henry Bigelow to devote a year to investigations on behalf of the committee and assembly of its report. After some initial doubts, he accepts the appointment at an annual salary of $7,500 and begins work on October 1.

Committee members visit Secretary of the Navy Charles Francis Adams to request a precise definition of the Navy's commitment to oceanography, and they keep in close touch with the Navy as Bigelow prepares their report. Adams will be an early WHOI board member, and his son, also named Charles Francis Adams, will chair the Institution's board forty years later.

1929

Bigelow casts a wide net as he solicits advice from scientists and laboratory directors around the world. He hires Virginia Walker to prepare the manuscript at a salary of $23 per week. Drafts and comments fly among the committee members. T. Wayland Vaughan plays a major role in the process, advocating a broad definition of oceanography. (Vaughan had overseen the name change from San Diego Marine Biological Institution to Scripps Institution of Oceanography a year after becoming director of the lab.) On January 26, Vaughan writes to Bigelow: "From the last paragraph in your [January 17] letter it appears to me that you are still looking on oceanography as an adjunct to biology. I would rather turn it around and look on marine biology as an adjunct to oceanography. You have been [led] to study the ocean primarily from a biological motive, whereas my motive was primarily a geological one. Since I am in the oceanographic game I should combat the consideration of the sea as an adjunct to geology no matter how important geology may be. . . . I think that the ocean should be studied as a thing for itself and as one of the most important parts of the earth."[20]

On November 18, Lillie presents the Committee on Oceanography's final report to the academy under the wordy title *Report on the Scope, Problems, and Economic Importance of Oceanography, on the Present Situation in America, and on the Handicaps to Development, with Suggested Remedies.* In the published version (simply entitled *Oceanography*), Lillie's introduction describes it as "an attempt to appraise the present condition of oceanographic research with reference to the more outstanding problems, so as to take bearings for future research. It is in no sense a textbook or a compendium of oceanographic knowledge." The journal *Science* publishes a summary of the report in January 1930 (see "A Developing View-Point in Oceanography," on pages 14–15). The report includes the committee's first specific recommendation: establishment of a well-equipped oceanographic institution on the Atlantic coast.*

Lillie acts quickly to ensure support for the East Coast laboratory. He notifies Max Mason, who has replaced Rose as president of the General Education Board, that NAS has endorsed the report. The Rockefeller Foundation had, in fact, authorized funding for the new oceanographic institution on November 14. Lillie also informally retains architect C. A. Coolidge to begin designing the buildings and wharves the institution will need.

Charles Francis Adams (second from right), Secretary of the Navy and one of the first WHOI trustees, poses with a group of naval officers while attending graduation exercises at the Naval War College in Newport, Rhode Island, in 1932.

---

* Other recommendations include the strengthening of three then-small laboratories in La Jolla, California; Seattle, Washington; and Bermuda. The Committee on Oceanography's work is not finished. It will continue, with Rockefeller support, for another eight years, appointing an array of subcommittees to study oceanographic problems. Bigelow succeeds Lillie as chair in 1935, when Lillie becomes president of the academy.

"A Developing View-Point in Oceanography"

Science, January 24, 1930

This article by Henry Bigelow discusses the Committee on Oceanography's "study of the scope, economic importance and present status of oceanography, with recommendations as to how this science may more effectively be encouraged in America."

To the general scientific public the most significant feature of the report is perhaps the general conclusion . . . that the establishment on our Atlantic coast of a new organization dedicated to the encouragement and prosecution of oceanographic investigation is 'the greatest need at the present time, both from the point of view of American oceanography and also for adequate participation of this country in a study necessarily international.' "

"The report is too long to be summarized here," Bigelow wrote. (A footnote offers it in "mimeographic form.") *"We, therefore, think it pertinent to set forth something of the view-point developed in the report that led to the recommendation of this particular kind of support for oceanography, rather than to the more conventional suggestion that our knowledge of the sea would be most rapidly increased by more deep-sea expeditions, and greater."*

Though the science of oceanography was born, he continued, *when first some fact about the sea was not only observed but recorded . . . it was not until about the middle of the nineteenth century that systematic examination even of the surface of the sea was seriously undertaken, or that scientists awoke to the fact that the underlying waters offered a whole new world for exploration. . . .*

Students of the history of science may well date the birth of modern oceanography from December 21, 1872, the day when the Challenger *set sail from Portsmouth, England, on her memorable voyage. And thenceforth, with every fresh venture below the surface of the sea, such a flood of new facts came pouring in that it seemed for a time as though this fact-catching could never lose its novelty. One great deep-sea expedition led to another, and more*

> *Most books* tell what their authors know or think they know. Henry B. Bigelow's "Oceanography" is devoted to telling what Mr. Bigelow does not know, to what, in fact, no one knows.
>
> —Lewis Gannett, *New York Times Book Review*, 1931

was learned about the sea during the last thirty years of the nineteenth century
than had been during the preceding three thousand.

When the quantity of "startling discoveries" gained from simple observation waned, the science began to stagnate.

And oceanography would probably be in a moribund state in America today,
just as the art of sailing a square-rigger is, but for the birth of the new idea that
what is really interesting in sea science is the fitting of these facts together,
and that enough facts had accumulated to make the time ripe for an attempt to
lift the veil that had obscured (and still obscures) any real understanding of the
marvelously complex and equally marvelously regulated cycle of events that
takes place within the sea. . . .

[I]n the further development of sea science the keynote must be physical, chemical and biological unity, not diversity. . . . [W]e believe that our ventures in oceanography will be most profitable if we regard the sea as dynamic, not as something static, and if we focus our attention on the cycle of life and energy as a whole in the sea, instead of confining our individual outlook to one or another restricted phase, whether it be biologic, physic, chemical or geologic. . . . [E]very one of us . . . must think and work in several disciplines . . . be either Jack of all trades or so closely in tune with colleagues working in other disciplines that all can pull together.

Science's synopsis of the report then briefly discusses some of the specific areas ripe for oceanographic study. Bigelow's full text was published by Houghton Mifflin Company in 1931 under the title *Oceanography: Its Scope, Problems, and Economic Importance.* In 263 pages, it describes oceanographic problems in terms still relevant nearly seventy-five years later.[21]

WHOI's first laboratory, later named for first director Henry Bigelow, opened for science in summer 1931.

1930—1939
The First Ten Years

1930 Articles of incorporation are filed in January, the board names Henry Bigelow director, and he initiates laboratory- and ship-building plans with funding from the Rockefeller Foundation

1931 The laboratory and coastal vessel *Asterias* are ready for the summer research season; *Atlantis*, with Columbus Iselin as master, arrives at the end of August

1932 A summer routine of bustling laboratories and weekly *Atlantis* cruises is established, while labs are largely bare in winter

1935 Maurice Ewing makes his first *Atlantis* cruise, using explosives for seismic work and initiating underwater sound as a mainstay of WHOI research

1936 Iselin publishes a major paper on the Gulf Stream, and WHOI research encompasses ocean biology, chemistry, physics, and geology

1937 Cooperative work with the U.S. Navy begins with investigation of the "afternoon effect" on sound in seawater

1938 Instrument development includes a current meter, the bathythermograph, and a multiple sea sampler

1939 Every room in the laboratory is filled for the first time with more than 100 summer researchers representing 29 institutions

1930

Articles of incorporation for the Woods Hole Oceanographic Institution (WHOI) are filed on January 6. The board of trustees, with members of the Committee on Oceanography as its nucleus, holds it first meeting on January 15 in New York City. The purpose of this new venture is "to prosecute the study of oceanography in all its branches; to maintain a laboratory or laboratories, together with boats and equipment and a school for instruction in oceanography and allied subjects."

The board elects Lillie as chair, Lawrason Riggs Jr. as treasurer, and Bigelow as director of the Institution and clerk of the corporation. The board members also hear a Rockefeller Foundation invitation to submit a formal request for funding.

Bigelow sets about recruiting scientists for the new institution. On January 22, noting that the Rockefeller funding is not yet available, he writes to Sverdrup, "All I can do at present is to ask in a perfectly informal way whether you would care at all to consider the position as Chief Physical Oceanographer at the new institution that we hope to see established?"[22] Though he will move to the United States in 1936 to be the third director of the Scripps Institution of Oceanography, Sverdrup is not yet ready to leave Norway. Bigelow invites Alfred Redfield, then teaching physiology at Harvard Medical School, to be senior biologist. After some negotiation, Redfield cables his acceptance from a sabbatical year in Europe. The following March, Bigelow notifies Redfield that he is recommended for a position as Professor of Physiology in Harvard's Zoological Department and that WHOI and Harvard will each pay half of his $8,000 salary. "Your decision to join us at Woods Hole

New York estate and trust lawyer Lawrason Riggs Jr., left, was MBL treasurer when WHOI was being organized. He was involved in the new institution's incorporation, helped draft its bylaws, and then served as treasurer of the corporation from 1930 to 1950, keeping all the financial records and guiding the investment program. Others in the photo taken on the WHOI pier are Columbus Iselin, Ben Leavitt, and Henry Stetson.

The *New York Herald Tribune* featured the new oceanographic institution in its magazine section on August 10, 1930.

has been a great relief to me," Bigelow writes, "for now I know that under your leadership marine physiology will be properly cared for—and I'm glad that we'll have the fun of working together."[23]

On February 13, the Rockefeller Foundation Executive Committee appropriates sums essentially identified by Lillie and Rose at least two years earlier for the founding and operation of the Woods Hole Oceanographic Institution. The foundation appears not yet to be affected by the stock market crash of October 1929.

Bigelow gathers ideas for a research vessel by consulting with anyone he can find who is knowledgeable about working at sea. His student Columbus Iselin, a seasoned sailor and Harvard's current Alexander Agassiz Fellow of Oceanography, is visiting European laboratories (at Plymouth, Paris, Monaco, Geneva, Berlin, Hamburg, Copenhagen, and Bergen). Iselin assembles a formidable plan for a research vessel while tending his wife, Nora, who has taken ill and spends three weeks in a Copenhagen hospital. "Lately I have been thinking a lot about boats of various kinds and after buying drawing instruments have been amusing myself designing 'the perfect' ocean research vessel," he writes to Bigelow on February 8. He encloses six handwritten pages of ship specifications, including staffing and costs, and a drawing, promising to hand-carry his deck layouts to Bigelow because they "are a little difficult to send across the ocean."[25]

In mid-February, letters between Bigelow and Iselin cross in the transatlantic mail. Bigelow's says, "I have it in mind that you are to have a position [at the new institution]," while Iselin writes,

With this 1930 sketch, Columbus Iselin thought he was drawing "a boat for Harvard."

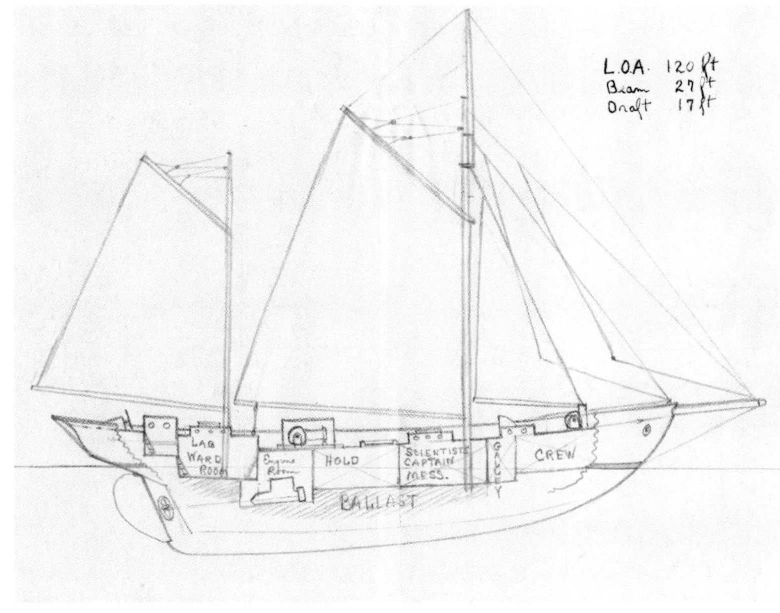

"It is amusing that I should have been thinking of very nearly the same boat as you have been. The long winded letter that I sent you was on a boat for Harvard. I never thought that the new institute would be so sensible as to build such a practical ship. . . . If you are looking for a captain for your new ship, I would like to have a chance at the job for a while. In some ways perhaps I am more qualified than the average professional."[27]

Much later, in a 1965 memoir, Iselin writes, "I made the first rough design and the firm of Owen and Minot of Boston translated this into a more modern and better engineered vessel."[28]

Columbus Iselin, left, and Henry Bigelow, center, chat with *Atlantis* captain Fred McMurray, facing Bigelow, and a visitor near *Atlantis* in the early 1930s.

A Seagoing Research Ship

From the time when the prospects of the new Institution were first discussed in earnest, the trustees have appreciated that one of the most important things that it could do would be to operate a seagoing research ship of moderate size, with convenient living quarters, and equipped to carry on investigation at all depths in various lines of sea science. No other American marine laboratory, independent of the government, is able to do this. A number of ships of various types were offered to the Institution, for purchase or as gifts. But none of them were satisfactory, either because of age, size, cost of upkeep, cost of conversion or because of other considerations.

The type of ship that would best fit our needs was the first question to be decided. Various scientific establishments in this country and abroad have most generously supplied us with plans and specifications of their ships, with full information as to their advantages and shortcomings. With this information at hand, and in view of the types of work that the Institution may be expected to carry out as well as the amount of money that can be allotted to the operation of the ship, it was decided that a steel auxiliary, of about 350 gross tons, about 140 feet long, with hull of sailing ship type, would best serve our purposes. The plans, as finally developed, call for an overall length of 142 feet, a beam of 29 feet, a draft of 16 feet, a ketch rig spreading 7500 square feet of canvas, and a Diesel engine of 250–300 H.P. to develop a speed of about 8–9 knots. Two laboratories are provided for, one on the boat deck and one on the main deck. There are comfortable accommodations in staterooms for six scientists, besides the officers, with dormitory space for several students, and ample accommodations for the crew. The main hoisting winch is to be located in the lower hold, but controlled from the deck. On completion of the plans and specifications, competitive bids were requested from reputable shipyards in the United States, Germany, Scotland, Denmark, Sweden and Norway. The bid made by Burmeister & Wain Ltd., of Copenhagen [$175,000] being substantially the lowest [the highest was nearly twice that], the contract was awarded to them.

—Henry Bigelow, WHOI's
first annual report, 1930

Designed by Boston naval architects Owen & Minot, *Atlantis* was built by Burmeister & Wain of Copenhagen. The keel was laid in late October 1930, and the ship was launched and christened on December 31 that year.

The Marine Biological Laboratory trustees are ready to transfer waterfront property to the new institution, but they determine that their charter requires that they receive fair market value for it. Purchase of land was not included in the Rockefeller Foundation appropriation, so the WHOI trustees apply to the Carnegie Corporation for $26,910 to cover the purchase price and, in the meantime, borrow $5,000 to secure the land. The title transfer takes place on June 23. "The trustees consider that the Institution has been particularly fortunate in obtaining this plot of about 33,594 square feet of ground," Bigelow notes in the first annual report, "for it combines the advantages of frontage upon deep water, a convenient location for a dock for our ship, an opportunity for expansion by further purchase in the future, and close vicinity to the laboratories already existing in Woods Hole."

By March 1930, Boston architects Coolidge, Shepley, Bulfinch and Abbott presented this concept for the first WHOI building. Construction began in July and the building was ready for occupancy the following summer.

The trustees formally engage the Boston architecture firm Coolidge, Shepley, Bulfinch and Abbott to design "a brick building, 135 feet long and 50 feet wide, of four stories and a basement, containing laboratories and research rooms to accommodate about 50 students and investigators, chart and reading rooms, refrigerating system, and the usual offices in the basement." * They approve the plans on May 3 and appropriate $400,000 for construction of the laboratory, dock, float, and seawall. Construction begins in July. (Actual cost, including furniture, will total $320,544.51.)

An original attraction of the Woods Hole site was common use of MBL's library, and the trustees appropriate $5,000 to contribute books and journal subscriptions to the library. Over the next decade, annual contributions from WHOI range from $600 to $1,000. Sharing of library facilities continues to this day.

---

* Unattributed quotes throughout the text are drawn from the published WHOI annual report for the year under discussion.

Henry Bryant Bigelow, First Director

Henry Bigelow was a rather quiet, modest, unassuming man, but in early-twentieth-century oceanography he was a giant. With incisive intelligence, gentle wit, and consideration, he brought out the best in others.

Michael Graham of the British Fisheries Laboratory at Lowestoft wrote, "One of his great qualities was that he had a definite effect on everybody who worked near him or dealt with him in any way. I remember . . . the impact he made on the International Council for the Exploration of the Sea when he came to see us in Europe in March 1931 as representative of the North American Committee on Fishery Investigation. Throughout the proceedings the conference was richer whenever he was present. . . . One had the feeling that he was a man of such excellence and such exceeding pleasantness that not for a moment would one relax in the effort to do one's very best in order to support him as far as possible." [29]

Bigelow's early career bridged the transition from the great exploring expeditions (his 1901 trip with Alexander Agassiz to the Maldive Islands was one of the last of these) to a "new" oceanography, seeking understanding of the "physical, chemical and biological unity, not diversity, for everything that takes place in the sea." [30] His definitive, late-1920s report for the National Academy of Sciences Committee on Oceanography (see pages 14–15) drew on his own experience in extensive studies of the Gulf of Maine and on voluminous correspondence with marine scientists and administrators around the world. He was, WHOI oceanographer Mary Sears said, perhaps "the last person [who] knew everything there was to know about oceanography." [31] In 1931, Bigelow was elected a member of the National Academy of Sciences and also received the academy's tenth Alexander Agassiz Medal in oceanography for his "outstanding contributions" to the field. (It is notable that he was only the second American to receive this medal; the first was Admiral C. D. Sigsbee.) After stepping down as director in December 1939, Bigelow served ten years as president of the WHOI board, was named chairman of the board in 1950, and held the honorary title of founder chairman from 1960 to 1967, when he retired from active participation in Institution affairs.

Born on October 3, 1879, into a well-to-do Massachusetts family that loved the out-of-doors, it was not surprising that young Henry became a naturalist. (His maternal grandfather, Henry Bryant, M.D., was an avid collector of birds' eggs.) Wilderness activities, including annual hunting and hiking trips, often in the company of his wife and children, provided a refuge, especially from the administrative responsibilities that he so disliked. He took up skiing in his late forties and continued it well into his eighties.

Henry Bigelow, Harvard professor and Woods Hole Oceanographic Institution director, 1930–1939.

Henry Bigelow thought everyone in the lab, including himself, should go to sea at least once each summer. Here he is at work aboard *Atlantis* in 1932.

Bigelow, on stern, and Iselin pull *Atlantis* away from the WHOI pier with Iselin's *Risk*.

Bigelow and his wife, Elizabeth, enjoyed a large circle of affectionate friends and colleagues, who shared their sadness when two of their four children died in their twenties—son Henry Jr. in a 1931 climbing accident and daughter Elizabeth of an embolism while horseback riding in 1934. Mary and Frederick survived their parents.

Henry Bigelow's primary professional home was always Harvard's Museum of Comparative Zoology, where he was an accomplished systematic zoologist and recognized authority on both the coelenterates (animals with internal body cavities, such as jellyfish) and fishes, particularly sharks, skates, and rays. His many awards included the National Academy of Sciences 1948 Giraud Elliot Medal for his later work on fishes of the western North Atlantic. Bigelow was the principal architect of the International Ice Patrol's hydrographic work in northern seas and coauthor of *Wind Waves at Sea: Breakers and Surf,* written at the invitation of the U.S. government for World War II operators of amphibious craft and other small vessels.

Henry Bigelow thought that a laboratory director's primary responsibilities were "keeping his staff happy, drawing in the best men to the Institution . . . and seeing to it that all hands are working in an environment favorable to productive research." [32]

Alfred Redfield affirms Bigelow's success according to that definition in the closing words of his contribution to the National Academy of Science's *Biographical Memoirs:* "His daily rounds of the laboratory at Woods Hole, in which he assessed our capabilities and kindly corrected our follies made him a great director. The proverbial uncle is an elder member of a family, detached from its petty turmoil, but kind and wise, to whom the youngsters turn for understanding and counsel. As such, he is remembered with affection and gratitude by his associates at the Museum of Comparative Zoology and the Woods Hole Oceanographic Institution, who knew him as Uncle Henry." [33]

Bigelow appoints Virginia Walker as secretary and administrative assistant. Over the next ten years, Walker (who becomes the wife of chemist Homer Smith in 1937) acts not only as business and personal secretary to Bigelow but also takes on the roles of purchasing agent, housing coordinator, and telephone operator, along with many other responsibilities.

Since there are no operating costs for 1930, at the suggestion of the Rockefeller Foundation, WHOI contributes much of the first year's funding to preparations for the 1931 Wilkins-Ellsworth Trans-Arctic Submarine Expedition. Using a rather aged U.S. submarine designated *0–12* but rechristened *Nautilus,* Australian explorer Sir George Hubert Wilkins plans to make submerged measurements under the ice and even make a run for the pole. These will be the first Arctic measurements since Norwegian Fridtjof Nansen's drift across the top of the world in the 1890s aboard *Fram.*

1931

Various difficulties and delays limit *Nautilus* operations. Loss of a dive plane on the crossing from Norway, where Sverdrup and others join the expedition, prevents cruising beneath the ice, but the scientific party does manage to record enough data to compose the first five contributions to WHOI's series of collected reprints.* These manuscripts will be presented in the March 1933 edition of *Papers in Physical Oceanography and Meteorology* (published jointly by WHOI and MIT). Their subjects include general oceanography of the area, echo-sounding data, gravity measurements, and bottom deposit descriptions. The latter is the work of the new WHOI research associate in submarine geology (and assistant curator of paleontology at Harvard's Museum of Comparative Zoology), Henry Stetson. The first submarine visit to the North Pole is delayed until 1958, aboard the first nuclear-powered submarine—also named *Nautilus.*

Iselin, at age twenty-six, travels to Copenhagen in April with his wife and baby to supervise the final stages of the construction of *Atlantis.* In a letter to Bigelow,

---

* Scientists at institutions like WHOI publish the results of their research in peer-reviewed journals. Upon publication, the journals make "reprints" of individual articles available for the authors to distribute. WHOI numbers these scientific contributions (they totaled more than 14,650 by December 2004) and gathers reprints of them annually into bound volumes. Through 1980, when copyright and other issues made it impractical, hundreds of copies of *Collected Reprints* volumes were distributed to libraries worldwide. Today the MBL/WHOI Library binds four sets each year to provide permanent records of Institution research.

The Burmeister & Wain shipyard in Copenhagen was best known for constructing steel freighters. Building the steel-hulled ketch *Atlantis*, shown here under construction at the yard, was a novelty for the firm.

The party aboard *Atlantis* for the Copenhagen sea trials included a number of nattily dressed men.

he reports that they have not taken the rooms Bigelow arranged for them because the 200-pound nurse "simply refused to carry the baby up and down so many flights of stairs." There are a million details to check, disputes to settle, and a narrowly avoided lockout at the shipyard. "My desk is simply covered with lists," Iselin writes to Bigelow on May 21. "Things are going remarkably well. I hope you are not too anxious about us. [First officer and former Cunardman Samuel] Clowser is such a very good man for this fitting out that it is nearly impossible for me to make bad mistakes."[34]

The shipbuilder and the architect are an unlikely alliance. Owen & Minot are famous as racing yacht designers, and Burmeister & Wain are best known for constructing steel freighters. Some of the difficulties encountered on *Atlantis*'s maiden voyage are no doubt due to a builder not used to accommodating downdraft from large sails and a designer not experienced with the drag of over-the-side oceanographic equipment. Iselin does his best to maintain a sense of humor. In one letter to Bigelow, he writes, "You will be glad to hear that Burmeister has fastened the mainmast track by two metal bands as well as the screws. They simply could not stand seeing the track left the way specified. They did this for nothing with my permission. I don't think Minot will climb up the mast to see if the track is more strongly fastened than specified."[35]

On July 2, *Atlantis* sails for Plymouth, England, with Iselin at the helm and a crew made up of his college friends; professional sailors from Denmark, Norway, and Finland; a short-order cook from Massachusetts; a British engineer; and Henry Bigelow's brother, eldest son, and his nephew Daniel Merriman, as well as Iselin's wife's cousin, the colorful Terry Keough, as bosun. Iselin reports that *Atlantis* sails too

"Atlantis" equipment account.
Paid in New York

Durable Wire Rope Company		$4.700
Richter & Weise - thermometers		1,000
Bergen Nautik - water-bottles		486
Eastern Ship Builders Inc., agents for –		
General Electric Company	$350	
Schindel McDaniels (bedding)	330	
M.N. Fogg (mattrasses)	560	
Rudmund Scofield (dishes cooking gear.)	700	
Hammacher-Schlemmer (tools)	200	
S. Appel uniforms	624	
Packing and freight	?	
	2764	— — — 2,800
Eastern Shipbuilding Corp Ltd – dorys		300
Kelvin & Wilfred O. White compass		100
Green – meteorological instruments		28
H. G. Butt – machinest		650
Jules Richard – meteorological instruments		60
Linnen thread and twine Co – 1 net		100 ✓
Hobbs & Warren Inc 1 Log book & printing		53.75
Hobbs & Warren Inc 1 meteorological log		60.
Ordered in Copenhagen by May 1st		$10,337.75
London tailor (officers uniforms)		$540
English outfitters (boiler suits & jerseys)		124
Henry Hughes & Son Ltd (chronometer, etc)		290
Negretti & Zambra – (Meteorological instruments)		325
C. F. Casella & Co Ltd - " "		100
Christiansen & Sorensen (nets)		400
		$1779

892.88 (marginal note beside H. G. Butt row)

Could copies of these bills be sent me? (marginal note, New York purchases)

Approximate in dollars of payments in kroner & pounds (marginal note, Copenhagen purchases)

I am keeping books with the price of each article, but when I left New York, the final bills had not come through for over half of the above. This was especially true of freight and packing charges. Have I forgotten anything that should be included in this account?

Iselin kept a careful accounting of equipment purchases for *Atlantis*. He sent this list to Bigelow on May 1, 1931.

The definitive sea trial for *Atlantis* took place in Copenhagen, on June 18, 1931, with Francis Minot representing naval architects Owen & Minot. The ship was accepted at a total cost of $208,129.77, plus $19,000 for oceanographic equipment purchased in Norway, Germany, and England. Though other problems would surface on the maiden voyage, WHOI officials were pleased to learn that the ship ran faster and consumed less fuel than promised.

much on her side, is "stiff," and pitches too much. "It takes two men ¾ of an hour each watch, hard work, to pump out the shit and water from the basins. The compressed air system does not work when the tank is on an angle and besides it uses up too much air. It would be too much of a job to put in a pump . . . here as the tank is hard to get at so we will simply have to pump our way across the Atlantic."

"Before leaving Copenhagen," he adds, "we gave a grand party. . . . [Danish oceanographer] Prof. [Johannes] Schmidt had the Prime Minister aboard and presented us with a flag."[36]

From left, crew members Knute Nielsen, John Churchill, Terry Keough, and Ona McClunen on deck during *Atlantis*'s maiden voyage.

Following shipyard work to correct problems encountered on the Atlantic crossing —including alterations to the plumbing system, the balky stove, and the rigging, plus installation of thirty-seven tons of additional ballast—*Atlantis* arrived in Woods Hole on August 31, 1931. Henry Bigelow (at left, wearing white trousers), was among those who gathered to welcome the new ship.

On July 16, with additional crew, the scientific party, and a final suite of equipment (winch, deep-sea thermometers, nets, water bottles, and chemical and meteorological apparatus) aboard, *Atlantis* heels gracefully to a west wind and heads for home.

The passage is not so graceful. The weather is mostly rough, the sewage problem gets worse before they sort it out, the stove doesn't work, everything leaks, equipment breaks, and there are several injuries, including a skull fracture for the chief engineer (not diagnosed until years later, following another bump on the head) and a leg injury for Iselin when he is smashed against a bollard during a squall. Nevertheless, in the grand oceanographic tradition, the scientists make physical, chemical, and light penetration measurements and collect biological samples as the ship zigzags across the Atlantic, arriving in Boston on August 26.

The Young Master

Marine historian Susan Schlee gives this history of *Atlantis*'s young first captain in her account of the ship's maiden voyage: "Lying sleepless in his pitching bunk, Iselin discovered that his cabin leaked."

> *Whatever may have crossed his mind that miserable night, it was probably not self-pity. He had been in worse situations at sea many times before. He had been on the water, in command of his own vessels, under all imaginable circumstances, for at least fifteen of his twenty-six years. He had grown up in New York City and New Rochelle (on Long Island Sound), and at the age of eleven had built his first leaky boat,* Sponge. *He had sailed and rowed his way through St. Mark's preparatory school, and as an undergraduate at Harvard University had bought a captured rum-running schooner,* Theresa White. *In the summer of 1925, he and some college friends made a trip in her to Nova Scotia, and the following summer, after graduation, he took possession of the first boat built for him, the seventy-seven-foot schooner* Chance. *With Terry Keough, Bart Hayes, and other friends, Iselin made an adventurous 5,000-mile voyage to the northern portion of Labrador to measure the temperature and salinity of the water and make other relatively simple oceanographic observations . . . encouraged . . . by his friend and teacher Henry Bigelow. . . .*
>
> *In the summer of 1927, Iselin, then enrolled in a master's degree program at Harvard, had made a more scientifically sophisticated expedition on* Chance *to study the Gulf Stream. He was having a still larger schooner built, and in 1928 sailed to Europe and back aboard his new* Atlantis, *bringing back water samples, plankton, temperature data, and, in spite of Prohibition, a barrel of port wine for a friend who was getting married. Iselin himself was married the following year, and in 1930 Bigelow asked him to join the Woods Hole Oceanographic Institution as master of the research vessel* Atlantis *and, more permanently, as a physical oceanographer. (Iselin, who had sold his 98-foot schooner* Atlantis *to Alexander Forbes, [later] a trustee of the Institution, asked Forbes if he would rechristen the schooner so that her name could be transferred. The schooner thus became* Ramah *and the Institution's ship* Atlantis.*)* [37]
>
> —Susan Schlee, *On Almost Any Wind*, 1978

Columbus Iselin aboard his schooner *Chance*, about 1926.

Iselin, at rail, and first mate Samuel Clowser donned proper uniforms for the arrival of *Atlantis* in Woods Hole on August 31, 1931.

---

* The second-choice name was *Penikese.*

Atlantis heads across the Atlantic on its maiden voyage.

In the summer of 1931 the steel-hulled ketch Atlantis *became the first deep-water research vessel for the new Woods Hole Oceanographic Institution on Cape Cod, and the serious investigation of the North Atlantic Ocean, which had come to a standstill some fifty years before, resumed. There had been a time in the nineteenth century when the study of the deep sea seemed to promise great rewards and several large oceanographic vessels had been engaged in research, but by the 1880s interest had declined, and by 1920 the last of what might be called first-generation research vessels had been scrapped or retired. Oceanographers were left without a means of studying the open ocean.*

With the arrival of Atlantis *a new round of investigations began and during the next thirty-five years the ketch made 299 cruises over more than a million and a half miles of ocean.* Because for six years she was the only vessel in the country large enough to undertake such extensive investigations, and because for another fifteen years she was one of only a few such ships, it was inevitable that scores of important discoveries were made from her decks. In fact, during her exceptionally long career, from 1931 to 1966, the ketch all but monopolized the testing of new instruments and the exploration of uncharted regions." [38]*

—Susan Schlee, *On Almost Any Wind,* 1978

---

* This mileage figure, thought to be accurate when Susan Schlee researched her book, was later recalculated. The correct number appears to be about 640,000 miles.

In early June 1931, Casey Boat Building Company of Fairhaven, Massachusetts, delivered the forty-foot (twelve-meter) launch *Asterias*. Named for a local starfish, and costing under $7,000, its purpose was to provide, as Bigelow wrote in the annual report, for "all the wants of the investigators at the laboratory during the . . . summer."

1932

The Woods Hole Oceanographic Institution commences its first full year of research. Staff members gather at Harvard to "divide up the pie." They appoint sixteen research assistants for a total of $4,000 in stipends. Bigelow's assistant, Mary Sears, who will complete her Ph.D. in 1933, is among them. The summer's staff, visiting investigators, and research assistants represent eighteen national and international institutions (including the New York Homeopathic College!).

A Well-Equipped Laboratory

WHOI follows MBL's practice of publishing an "annual announcement" to entice summer researchers. "There are few oceanographic problems but can be attacked profitably at Woods Hole," the 1932 edition says, and describes scientific programs planned for the year as well as facilities available to pursue them. Details were offered regarding the laboratory building, first occupied on June 15, 1931:

> The main building is a four-story brick and concrete structure 136 ft. long by 50 ft. deep, of the simple type of construction usual in modern laboratories. In the basement are the receiving and shipping rooms, boiler room, battery and transformer rooms, the storeroom for chemical and other apparatus, a room in

which constant temperature can be maintained, a refrigeration room, and one laboratory containing concrete aquaria, some of which are piped with chilled as well as unchilled sea water. There is also a machine shop, for the repair and construction of apparatus used in the laboratory and on the ship.

The first floor contains the offices, the director's room, a large chemical laboratory and nine smaller research laboratories. On the upper floors are the reading room, chart room, camera and drafting room, two dark rooms for experimental work, one camera dark room and twenty-three research laboratories, one of which is fitted as an aquarium room. Eight of the research laboratories, in addition to the one large chemical laboratory, are provided with fume hoods. Most of the rooms have salt water tables of the type now widely used in marine biological laboratories, while other rooms are designated for physical investigations. Sixteen of the laboratories are designed for individual use, the others for the use of groups of two or more investigators. The rooms are simply but adequately fitted with tables, counters, drawers and the usual movable furniture. Each is provided with a sink with fresh water (in addition to the salt water tables just mentioned), with gas, and with electric outlets for power as well as for light. Adequate heating is provided for winter occupancy, and it is planned to keep the laboratory in operation the year round.[39]

Bigelow's U.S. Bureau of Fisheries comrade in fish investigations, William C. Schroeder, is appointed business manager in May—his father, also William but better known as "Pop," a plumber by trade, becomes superintendent of buildings and grounds, and his mother makes plankton nets for WHOI researchers. On June 1, Bigelow appoints Captain Frederick S. McMurray temporary master of *Atlantis* so that Iselin can focus on physical oceanography. This arrangement becomes permanent on October 10. Iselin continues to supervise ship operations. However, throughout the 1930s, Iselin plays a strong role in Institution business (his first title is general assistant to the director and master, R/V *Atlantis*). He lives on Martha's Vineyard in the wintertime, generally spending Mondays and Fridays at WHOI and the rest of the workweek in Cambridge, where Bigelow is at work in his office at the Museum of Comparative Zoology. Iselin commutes to the Woods Hole laboratory in his thirty-eight-foot (eleven-and-a-half-meter) boat *Risk,* which will make some ten thousand crossings of the sound during his WHOI career. "The first winter I was at Woods Hole," Iselin will tell one interviewer, "I was the only person in the building except the janitor. I came in one morning in a northwest gale. I got off my boat, went into the office and put my pants on the radiator to dry. That was the day A. Lawrence Lowell, the president of Harvard, walked in for a visit."[40]

It took many hands to hoist *Atlantis's* sails, here the jib.

In July, *Atlantis* makes two short cruises to continue U.S. Bureau of Fisheries investigations of mackerel biology when funding is cut for the bureau's vessel *Albatross II.* Among other additions to *Atlantis's* scientific equipment, a "sonic sounding machine" is installed in October. By the end of December, 422 soundings have been taken in depths greater than 660 feet (200 meters). Through the 1930s, similar improvements and diligent maintenance keep the ship available to scientists.

Alfred Redfield was fond of telling a story about how he learned the definition of a sailor on the first day of his first *Atlantis* cruise: *A cold front [came] by during the night and the next morning there was a pretty good northwest breeze and the sea was pretty bobbely [sic]. I came up on deck, hung on to something, and looked around. . . . I made my way aft and there I saw the ship's carpenter in the scuppers, caulking the deck. As I watched, he laid down his caulking mallet, removed his dentures, put them in a safe place, leaned over the rail and put his breakfast. Then he came back, replaced the dentures, picked up the mallet and went right on with his job. I learned then the way an oceanographer behaves.*

—From a transcript of Alfred Redfield's anniversary address in celebration of the fortieth anniversary of the Woods Hole Oceanographic Institution, 1970

In September, commanding officer Charles G. Slayton of USS *Hannibal* is the first U.S. naval officer to join *Atlantis* for training in the use of oceanographic equipment.

1933

Atlantis statistics for 1933 include 198 days at sea (largely in the Gulf of Maine and the Caribbean Sea), sailing 17,605 miles (28,168 kilometers), making 5,357 temperature and salinity observations, and taking 18 mud cores, as well as 378 phytoplankton and 135 bottom samples.

Summer fellows include Ancel B. Keyes, who received the first Ph.D. in oceanography conferred in the United States, from the Scripps Institution of Oceanography in 1930. His summer work focuses on ammonia in the sea.

The 1933 annual report lists the Institution's first thirty-seven publications, encompassing topics in biology, geology, geophysics, physical oceanography, microbiology, and chemistry. (In 2004, WHOI investigators will publish more than 450 scientific papers.)

1934

From its launch in 1931 until December 31, 1934, *Atlantis* is at sea 646 days, traveling 60,719 miles (97,150 kilometers) and spending about half the time under sail alone and only a small part of the time under power alone. The ship's costs for fiscal year 1934–35 (exclusive of insurance and new construction) total $38,411. (By comparison, operation of three large research vessels will cost more than $16 million in 2004.) New construction includes a dory and two whaleboats, a pair of davits, and reconstruction of the cabin skylight "to remedy structural weaknesses." The ship also is outfitted with a new steering wheel, new washbasins, and running salt water in the upper lab.

In the summer of 1932 *the* Atlantis *was at last in Woods Hole, and the routine that would become characteristic of the institution's prewar years was established. Throughout July and August the* Atlantis *went out on weekly cruises [women on day cruises only], and every student and professor in the laboratory, regardless of the delicacy of his or her digestion, was expected to go out at least once during the summer. Bigelow wandered through the laboratory helping Stetson to organize the* Nautilus' *bottom samples, stringing a plankton net on a frame for a surprised lady graduate student, or squinting through a microscope at the plankton that Alfred Redfield had collected.*[41]

—Susan Schlee, *The Edge of an Unfamiliar World,* 1973

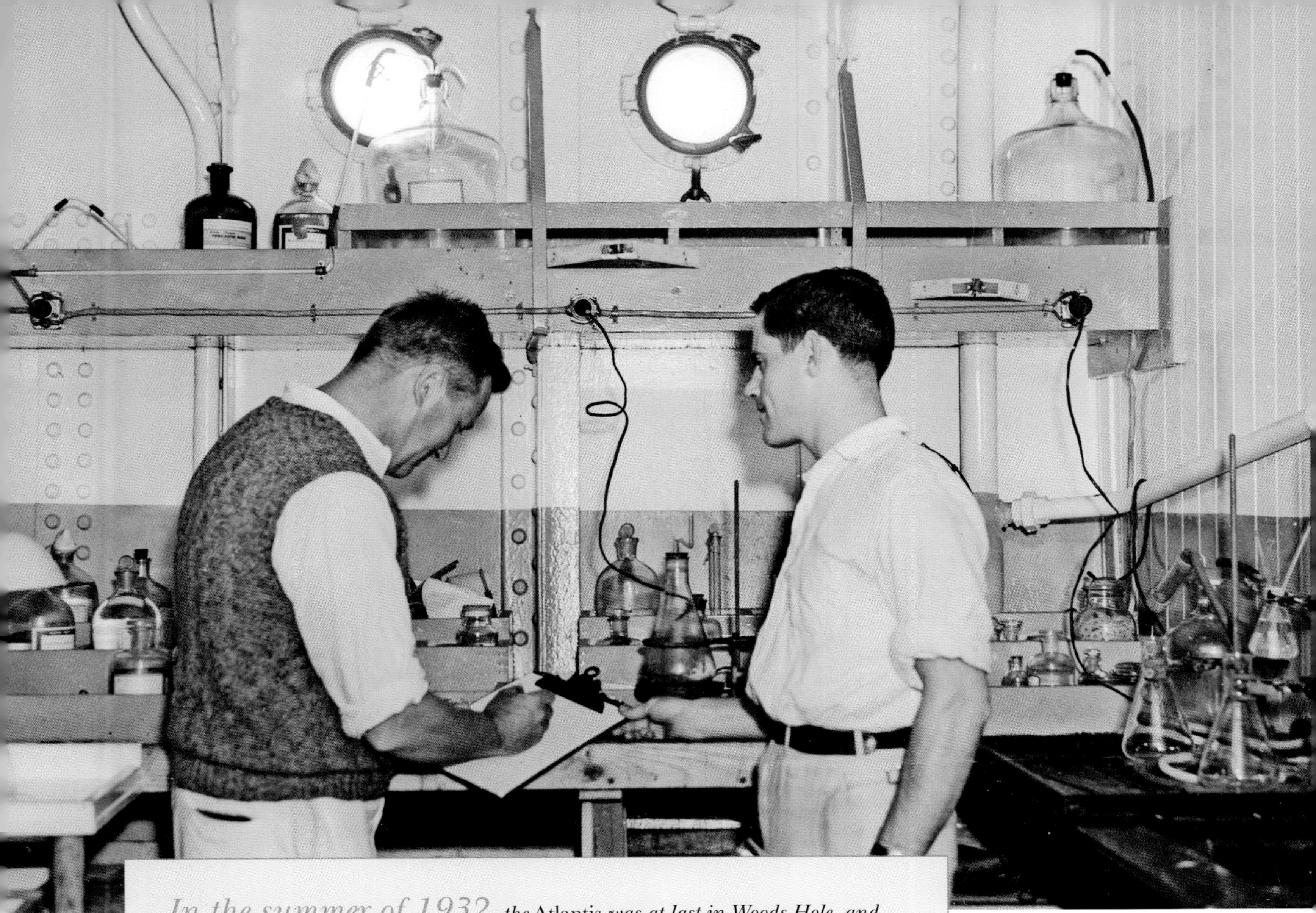

Alfred Redfield, left, and Homer Smith set up their chemical apparatus in the lower lab of *Atlantis* in 1933.

Atlantis offered few amenities. Here a shirt gets a scrub on the ship's deck.

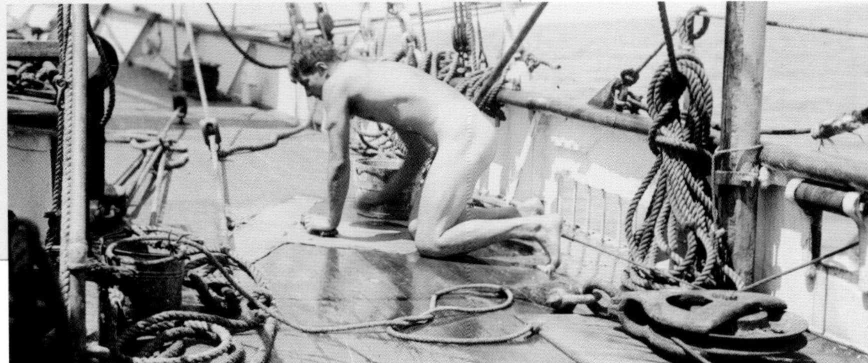

1935

A fatal car accident in Pensacola, Florida, takes the lives of two *Atlantis* crew members, first mate William Delano Potter and radio operator Spencer Greenwood. In 1936, Potter's mother will contribute a handsome carved walnut bookcase for the *Atlantis* wardroom in her son's memory.

As the result of changes in policy—and feeling the effects of the Depression—on April 10, 1935, the Rockefeller Foundation trustees vote to appropriate $1 million in endowment to the Woods Hole Oceanographic Institution in place of the remaining five years of promised operating support, noting that this should "make the Institution permanently independent of outside sources."[42]

At the end of the fiscal year, the "retiring allowance" fund established in June 1934 has seventeen voluntary participants and contains $6,197.75 ($3,050.46

from salary deductions, $3,050.45 contributed by the Institution, and $98.64 in interest from the savings bank).

In an early ocean-observatory effort, Harvard biologist (and WHOI "junior biologist") George Clarke, who sailed on *Atlantis*'s maiden voyage, establishes a station seven miles off Woods Hole. At this location "as far removed from land influences as possible and still accessible from Woods Hole," he makes monthly observations of a sustained population of the copepod *Calanus finmarchicus* and its changes in relation to tides and seasons. The work includes net and pump collections, temperature and light penetration measurements, and water sampling and analysis.

On an October *Atlantis* cruise, after allaying Bigelow's fears for the ship, Maurice Ewing of Lehigh University tries out his new method for employing dynamite as a wide-spectrum sound source to determine both the topography and structure of the ocean bottom by seismic reflection and refraction. The cruise objective is to measure the thickness of continental shelf sediments near the entrance to Chesapeake Bay. On another cruise to Georges Bank, Charles Piggot of the Carnegie Institution's Geophysical Laboratory tries out a new coring tube that is operated by explosives. Some of the cores he recovers are more than 8 feet (2.4 meters) long and are among the longest taken to date. Over the years, improvements in coring technology will allow increasingly longer cores to be collected. In 2005, WHOI researchers were constructing a new device capable of bringing home cores as long as 132 feet (40 meters).

Explosive charges set off near the surface from *Atlantis*'s whaleboat provided the sound source for Maurice Ewing's pioneering work using a seismograph, which recorded the sound reflected from the seafloor and the sediment layers beneath it. Twenty years later, Brackett Hersey's underwater sound group was still using the same approach; this photograph was taken in 1955.

Maurice Ewing—A Lasting Impression

Maurice Ewing's first step aboard *Atlantis* in the summer of 1935 initiated a long and productive association. He became a valued associate of WHOI directors and staff, set a fierce pace for research at sea, initiated underwater sound work whose legacy continues at WHOI today, and brought the Institution one of its most creative long-time staff members in Allyn Vine.

In late 1934, WHOI trustee William Bowie, chief of the Division of Geodesy for the Coast and Geodetic Survey (CGS), visited Ewing at Lehigh University, where he was a physics instructor. Bowie offered funding and ship time to determine whether the seismic method for prospecting on land could be used to investigate the structure of

Maurice Ewing holds a release and its timing mechanism—a block of salt—for seismic equipment that is to be lowered from *Atlantis*.

the continental shelf. Previously, Ewing's advocacy for using geophysics to study the earth had found little support. "If they had asked me to put seismographs on the moon instead of the bottom of the ocean," Ewing said later, "I'd have agreed, I was so desperate for a chance to do research." [43]

When the CGS ship *Oceanographer* (formerly H. P. Morgan's yacht *Corsair*) proved a difficult working platform, Bowie arranged for transfer of the work to *Atlantis* in the fall of 1935. Over the next few years, during a few two-week *Atlantis* cruises, as well as in land-based excursions to the Pennsylvania countryside, Ewing and his students designed and built their own instruments and developed seismic techniques.

At sea, they first used an *Atlantis* lifeboat as a shooting ship, with the receiver resting on the seafloor at the end of the anchor cable. (After dark, to the surprise and consternation of the ship's crew, the energetic Ewing floated charges off the stern of the ship for remote detonation.) The group later tried stringing several receivers and charges along a cable and, later still, free-falling gear with explosives set off by attached clocks. These techniques proved too complex to work well, but Ewing clung tenaciously to his belief that seismic refraction lines could be run in the deep sea. He was right, but success with a simpler approach was delayed until after World War II.

Ewing and his student Allyn Vine also developed early deep-sea camera techniques. In March 1940 they produced the first photographs of the deep seafloor. "[T]hey gave a cloudy view of a sandy ocean bottom with sketchy shadows that probably indicated rip-

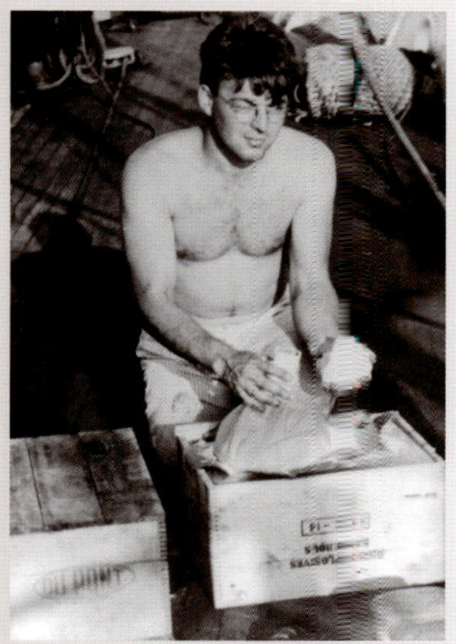

ple marks. Later Ewing demonstrated that such marks were common in deep water, a surprise to most geologists, who had considered the deep sea devoid of currents strong enough to produce marks of any kind."[44] Both the photography and the seismic experiments involved pioneering work in creating watertight equipment for use in deep water.

During the war, projects undertaken by Ewing's group included improvement of the bathythermograph (see pages 56–58) for efficient use from ships and on submarines, but their best-known efforts were perhaps the discovery and exploitation of the ocean's low-velocity sound channel. The sofar (*sound fixing and ranging*) channel, occurring at depths of 2,300 to 4,300 feet (700 to 1,300 meters), may be described as a pipe whose boundaries repeatedly reflect sound or as a wave-guide that traps sound waves. Thus an explosion near the depth of minimum sound velocity spreads in two dimensions instead of three and can reach great distances before its signal is lost in ambient noise: a few-pound charge dropped off the west coast of Africa in one of Ewing's experiments could be heard off the Bahamas. Applications of this discovery have included listening stations for locating downed pilots, studies of ocean phenomena that can be detected using sound, and tracking of current-following floats.

When Ewing moved to Columbia University in 1946, he left a lasting impression on the young Woods Hole Oceanographic Institution. British oceanographer Edward Bullard wrote of him, "He was driven by an inner urge to compulsive overwork. He believed that every opportunity must be seized and exploited to the full. . . . [His] style of work had an electrifying effect on what had been a rather slow-moving marine biological station. In his unpublished memoirs Iselin wrote: 'He had a profound effect on the success of this laboratory. He arrived here first as a very young professor. . . . He brought with him several Lehigh students and the place has never been the same since. They literally worked night and day, and seven days a week.'"[45]

Al Vine mixes explosives aboard *Atlantis* in the late 1930s. The TNT was melted in a steam cooker on the port side and poured into lengths of steel pipe on the starboard side. A clock was affixed to each bomb before it was launched.

Maurice Ewing, left, consults with *Atlantis* captain Fred McMurray during a cruise in the late 1930s. Twenty years later, McMurray was instrumental in saving Ewing's life when Ewing, his brother John, and two others were washed overboard while trying to secure gear in a gale aboard the Columbia University vessel *Vema*.

1936

A paper by Iselin summarizes several years of work on North Atlantic circulation and provides the first good concept of the general setting and environment of the Gulf Stream.[46] Iselin attends the International Geophysical Union meeting in Edinburgh and also represents WHOI in London at conferences concerning a proposed joint research program with the Bermuda Biological Station. In Copenhagen, he discusses repairs to *Atlantis*'s fuel pumps with Burmeister & Wain.

1937

Cooperative work with the Bermuda Biological Station commences with *Atlantis* Cruise 66 in June, Cruise 72 in October, and Cruise 73 in December. WHOI provides scientific oversight for these studies of "whether or not oceanic circulation changes sufficiently from year to year to be an important factor either in the fisheries on the two sides of the Atlantic or in the climatic variations in northern Europe."

WHOI Science in 1937

Edmund Watson deploys his electric current meter from *Atlantis,* about 1938. Al Woodcock is at right.

A list of forty-seven projects in the 1937 annual report illustrates the breadth of scientific work under way at WHOI in the 1930s. It encompasses physical oceanography and meteorology, physical and biological chemistry of seawater and marine sediments, physiology of marine organisms, plankton studies, marine microbiology, and submarine geology. The projects and investigators involved include the following:

- Labrador Sea physical oceanography and International Ice Patrol reports (Edward Smith, Floyd Soule)
- Eastern slope water hydrography (Columbus Iselin, Alfred Woodcock, Kurt Kraus)
- Equipment for measuring and studying light intensity at various depths and studies of diurnal plankton migration (George Clarke, W.R. Sawyer)
- Electric current meter design (Edmund Watson)
- Theoretical and experimental investigation of rotation and stratification in wake streams (Carl-Gustav Rossby, Athelstan Spilhaus)
- Various studies of oxygen and nitrogen in seawater, marine organic matter, and sediments (Victor Emmel, Theodor Von Brand, Norris Rakestraw, Charles Renn, Selman Waksman)

- Bacteriology and biochemistry of ocean muds (Selman Waksman, Margaret Hotchkiss, Cornelia Carey, Yvette Yardman Unto Variovaara)
- Distribution of phosphorus and ammonia in the Gulf of Maine (Alfred Redfield, Homer Smith)
- Copepod nutrition (George Clarke, David Bonnet)
- Growth rates in cod scales at different temperatures (William Schroeder)
- Color changes in fishes (G. H. Parker, Alexander Abramowitz)
- Phosphate and nitrate assimilation in diatom photosynthesis (Bostwick Ketchum)
- Testing of various zooplankton nets to determine the smallest size suitable for quantitative investigations (George Clarke, Dean Bumpus)
- Identification of Gulf of Maine diatoms collected from *Atlantis* (Lois Lillick)
- Seasonal and regional distribution of planktonic animals over the continental shelf from Cape Cod to Chesapeake Bay (Henry Bigelow, Mary Sears)
- Study of medusae collected by William Beebe at Bermuda in 1929–30 (Henry Bigelow)
- Foraminiferal indications of past temperature off the continental slope (Fred Phleger)
- Development of seafloor seismic instruments (Maurice Ewing)
- Geological significance of submarine canyons found at the edge of the continental shelf between Georges Bank and Delaware Bay (Henry Stetson)

Young WHOI scientists Rudolph Nunnemacher, left, Charles Renn, and Mary Sears take a break in the late 1930s.

Before wartime expansion, WHOI facilities (at right) consisted of the laboratory, the pier for *Atlantis*, the small boat basin behind the pier, and a small pump house/utility building near the pier. When this photo was taken, about 1940, the Penzance Garage—the long building to the right of the WHOI facilities, built about 1924 to provide a place for chauffeurs to service the cars of Penzance Point residents—was owned by the MBL. The buildings at far left housed U.S. Bureau of Fisheries laboratories and dormitories, and MBL facilities and private residences are located between WHOI and the Fisheries property.

Early in 1937, with Iselin as chief scientist, *Atlantis* joins the navy destroyer *Semmes* off Cuba for investigations of the "afternoon effect." Thus begins WHOI's operational cooperation with the military. The objective is to determine why the *Semmes* sonar system works only part of the time. After two weeks of exercises, Iselin concludes that the temperature structure of the water is causing the problem. For example, on a hot, still afternoon, the top several hundred feet of water comprise distinct layers of warm and progressively cooler water that bend sonar beams much as a prism bends light passing through it. These distortions make it very difficult for *Semmes* to track a submarine.

1938–1939

During cooperative work in Cuban waters with University of Havana colleagues aboard *Atlantis,* Bill Schroeder collects more than a thousand species of fish and invertebrates, including about one hundred previously unknown species from depths of 660 to 9,500 feet (200 to 2,886 meters).

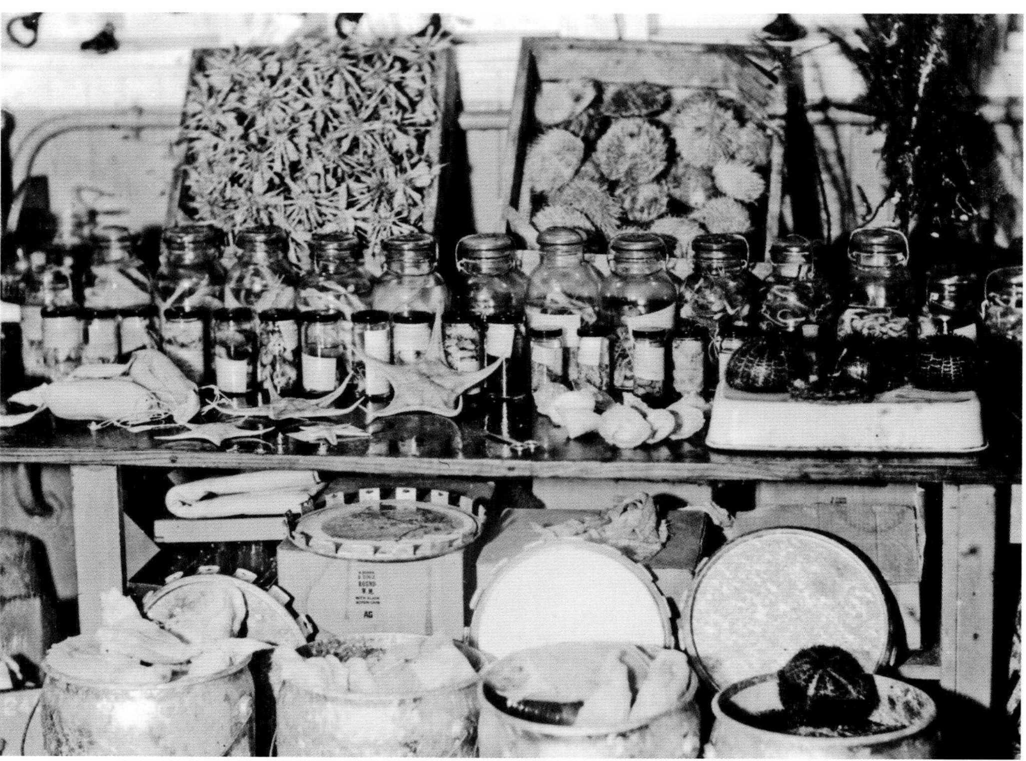

Samples collected near Cuba are neatly arrayed for a portrait aboard *Atlantis.*

The 1938 annual report comments on instrument development: "The recording current meter which was built last summer under Professor Watson's supervision was tested, successfully, on *Atlantis* to a depth of 250 meters. And the automatic recording apparatus, which goes on the ship, was perfected.

The End of an Era

On December 16, 1938, Henry Bigelow wrote to Norwegian colleague Johan Hjort, *From the personal standpoint my chief news is that last August I announced to the Trustees my desire to resign next year from the directorship of Woods Hole Oceanographic Institution. I will then be 60 years old, and I really don't want to continue any longer. Redfield, Conklin and Vaughan were appointed a committee to consider the matter of a successor. I rather hope and expect that it will be Iselin, but please don't say anything about this yet. It will be a great relief for me to drop out of administrative affairs, which I hate. And, as I'll keep my Harvard professorship, I will have plenty left to do, with teaching in the university, and research.*[47]

The Water Street drawbridge was carried away by a high tide during the hurricane of 1938. It was more than a year before cars could use the street again, but a temporary arched bridge accommodated foot traffic. *Atlantis* arrived in homeport just before the hurricane ("we made 15 knots…under sail coming in"[48]) to ride out the storm safely tied to a pair of moorings off the pier. Pop Schroeder's vigorous caulking kept the laboratory basement largely dry so that damage to WHOI property was minor. Repairs included resetting stone blocks in the sea wall, reconditioning lawns, and resurfacing the driveway. (And Dean Bumpus's wallet, washed out of his pocket while he was seated on a bar stool at Water Street's Cap'n Kidd during the storm surge, was found in a nearby rosebush the next spring.)

"Professor Spilhaus has likewise developed two instruments for the purpose of speeding up observations in the surface layers of the ocean; namely a Bathythermograph, and a pressure operated Multiple Sea Sampler. The rapid technique made possible by these instruments will, it is hoped, make a closer approach to synoptic observations practicable."

For the first time, every room in the laboratory is in service in summer 1939. More than one hundred individuals representing twenty-nine institutions

(thirteen more than the previous year) work in WHOI labs or aboard *Atlantis*. Institutions represented range from Harvard, Yale, and Southwest Missouri State Teachers College to the Fouke Fur and the Submarine Signal companies, Cushman Laboratory for Foraminiferal Research, and Glasgow's Royal Technical College.

On September 28, 1939, with security tightening along the U.S. coastline, Bigelow takes the unusual step of sending the location and objectives of October and November *Atlantis* cruises to the secretary of the navy.[49]

A short, unpublished history of the Institution written by Iselin in about 1960 comments, "By the end of ten years we had made considerable progress in understanding the currents of the western North Atlantic. In short, after the *Atlantis* had cruised for ten years, we felt generally at home in most of the North Atlantic, both physically and chemically, and to a lesser extent geologically and biologically."

On December 31, 1939, Henry Bigelow steps down as director, passing the mantle to Columbus Iselin. Bigelow succeeds Lillie as president of the Corporation.

The Atlantic's Mighty Meandering River:
A Short History of WHOI Gulf Stream Research

The Gulf Stream is the major current system closest to WHOI, and it has been an important subject of study throughout the Institution's history. Columbus Iselin described the circulation of the western North Atlantic in a 1936 paper[50] based on extensive but widely spaced temperature and salinity measurements from *Atlantis,* using an array of reversing thermometers and sample bottles. Over time, more capable ships, better instrumentation, and the determination of those willing to spend long, hard days at sea produced an ever-clearer picture of the stream itself, related features, and its place in global circulation.

World War II brought two significant advances to this work—more accurate navigation in the form of loran (based on radio signals sent out by shore-based stations) and the bathythermograph BT—see pages 56–58). In the postwar period, scientists could detect the finer structure of the stream, its twists and turns, narrowness and intensity, and eddies and meanders.

During the three-week, six-ship Operation Cabot survey in 1950, physical oceanographers both collected extensive data and, for the first time, followed the

Atlantis was photographed in rolling seas from the Canadian vessel *New Liskeard* in June 1950 during a six-ship survey of the Gulf Stream called Operation Cabot, the largest survey of the stream to that date.

evolution of a cold core eddy as it moved southward and pinched off from the Gulf Stream. The following four decades would bring intense scrutiny of both cold core (south of the stream) and warm core (north of the stream) eddies or "rings," revealing that ring behavior includes splitting into pieces, merging with one another, interacting with the Gulf Stream, reforming as modified rings, and coalescing completely with the stream.

In the 1950s, a combination of new theories, neutrally buoyant floats ballasted to follow deep-water motions, extension of hydrocasts to the deep sea, and rotating tank experiments revealed the Gulf Stream's expression down to the very bottom of the water column as well as the deep western

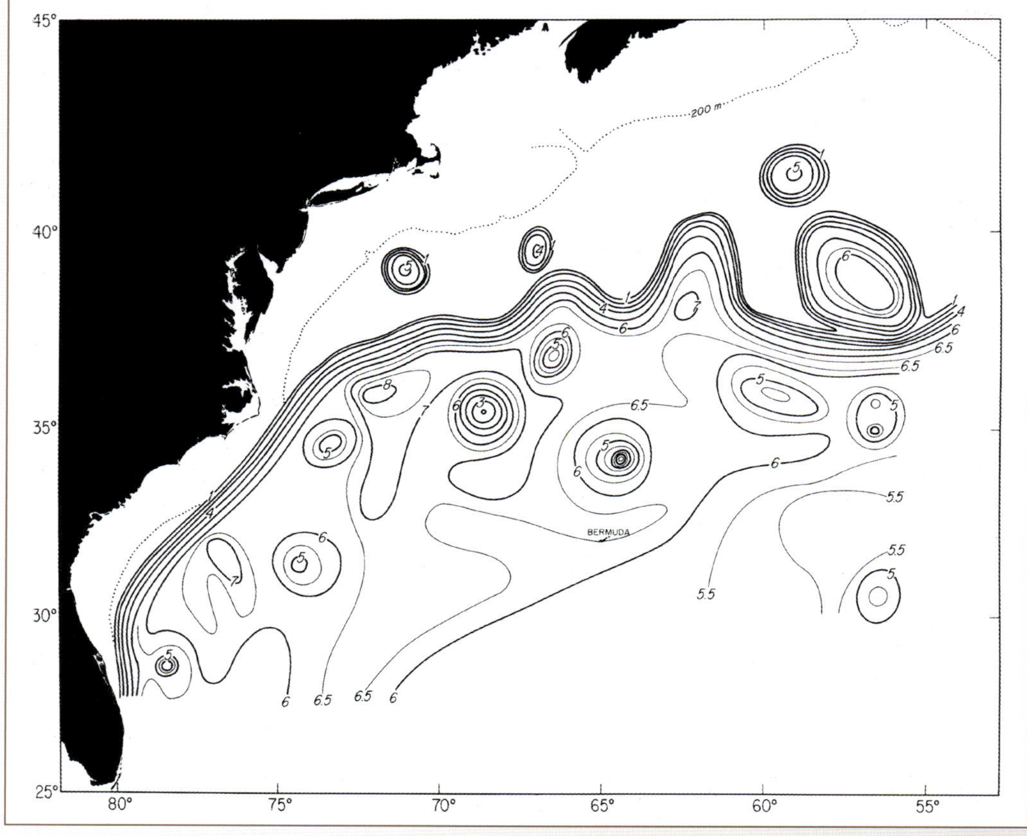

This chart of 1975 hydrographic and satellite data shows the path of the Gulf Stream and a number of warm-core and cold-core rings (north and south of the stream, respectively) that have spun off it. This was the first attempt to create a "weather map" for the northwestern Atlantic Ocean. The flow is charted using the depth (in hundreds of meters) of a water layer registering 59°F (15°C), which years of research have shown to be a reliable indicator of the surface current.

countercurrent that flows south sometimes below and sometimes near the stream.

The next instrumentation step needed was a means to make long-term measurements, a very challenging task due to the variability of the swiftly flowing currents. Experimentation with moorings and current meters began in the late 1950s, and it took ten to fifteen years of unremitting effort for the WHOI Buoy Group to develop a trustworthy system for routine use. The moored current meters brought reliable estimates of mean current flow in the vicinity of and under the Gulf Stream and revealed fluctuations with time and variations in speed of flow.

Radio- and later satellite-tracked surface drifters brought new definition to the stream's upper flows in the 1960s and 1970s. During the same period, early subsurface floats that required a following ship to chart their progress gave way to long-lived sofar floats tracked through moored listening stations. Both provided the means for more detailed understanding of Gulf Stream fluctuations, meanders, and rings.

The late twentieth century brought continued refinement of techniques and synthesis of Gulf Stream data. Today, WHOI scientists and their colleagues apply their long experience with investigations of the western North Atlantic to circulation studies worldwide. They employ expendable BTs, conductivity-temperature-depth profilers, satellite navigation, and acoustic Doppler current profilers—the modern versions of the BT, hydrostation, loran, and early current meters—as well as advanced profiling floats and other state-of-the-art tools such as satellite altimetry and infrared imaging.

An electronic current meter goes over the side of *Crawford* in 1965. Development of modern, long-lived current meters was vital to progress in Gulf Stream research.

This feature is based on an unpublished 2004 manuscript by WHOI scientist emeritus Philip L. Richardson, entitled "WHOI and the Gulf Stream."

Probing the Seafloor: A Short History of WHOI Marine Geology and Geophysics

In 1930, the depths of the main ocean basins were known only approximately from widely spaced soundings made with lead-weighted lines and, after World I, with an early echo sounder. Geologist Henry Stetson, with a joint appointment at WHOI and Harvard's Museum of Comparative Zoology, began to dredge rocks and core sediments from nearby continental margins during *Atlantis* Cruise 34 in 1934. The following year, Maurice Ewing of Lehigh University used *Atlantis* as a platform for adapting continental geophysical techniques to seafloor studies. Stetson's knowledge of sediment characteristics and Ewing's knowledge of the behavior of sound in seawater and the seafloor were both put to lifesaving use during World War II for submarine detection and defense.

After the war, Stetson continued his extensive collection of seafloor samples, taking *Atlantis* into the Pacific for the first time in 1955 on what was, unfortunately, his last cruise (he died at sea off Chile). Collection and analysis of seafloor rock and sediment has continued to be an important WHOI objective since then, and the Seafloor Samples Laboratory today houses more than fourteen thousand marine geological samples collected by coring, dredging, and other means. WHOI scientific and technical staff members have been key

Henry Stetson, assistant curator of paleontology at Harvard's Museum of Comparative Zoology, was among the earliest appointments to the WHOI staff. As research associate in submarine geology, he designed early coring and dredging equipment. He stands next to the top section of the Hvorsler-Stetson corer.

The chain-bag dredge, shown bringing its cargo aboard *Atlantis,* has been an important tool for marine biologists and geologists. Maurice Ewing is at left.

players in the progress of coring techniques in recent decades, and they are currently developing a new large-diameter coring device for retrieving sediment cores as long as 132 feet (40 meters).

Though Ewing moved from WHOI to Columbia University in 1946, he continued to use Institution ships for several years. During a seismic refraction experiment conducted from *Atlantis* and *Caryn,* he, Brackett Hersey (who had studied with Ewing for his Ph.D. degree), and other colleagues gathered firm evidence showing oceanic crust to be fundamentally different (thinner) from continental crust. Hersey was in charge of WHOI geophysics research through the mid-1960s, building a multidisciplinary group that grew to about fifty people who pioneered many new techniques and instruments, including the continuous seismic profiler for visualizing and quantifying layered sediment beneath the seafloor.

Bigger, more capable ships and improved instruments brought further progress in marine geology and geophysics, with WHOI scientists collecting data and contributing ideas to the diffuse, international dawning of the earth science revolution known as plate tectonics in the late 1960s. Today scientists in the Geology and Geophysics Department employ descendants of the early techniques and tools, as well as remotely operated and autonomous vehicles and long-term seafloor recording devices not dreamed of in the 1930s, as they advance a wide range of marine earth sciences research. They engage in a broad spectrum of studies—sedimentology, paleoceanography, seismology, geomagnetism, electromagnetics, tectonics, petrology, geochemistry, high-pressure physical properties, geomicrobiology—all of which promise a strong role for the Woods Hole Oceanographic Institution in the future of seafloor research.

Brackett Hersey test electronic gear in the upper lab aboard *Atlantis*. He joined the WHOI staff in 1947 to run an underwater acoustics program and later became the first chairman of the Geology and Geophysics Department. Navy funding for his group played a major role in sustaining the Institution in the uncertain postwar period. After moving to the Office of Naval Research in 1966, Hersey maintained close ties at WHOI.

This feature is based on an unpublished 2004 manuscript by WHOI scientist emeritus Richard P. Von Herzen, entitled "A Brief History of Marine Geology and Geophysics at WHOI."

A crowd gathers to send *Atlantis* off on a Mid-Atlantic Ridge research voyage in 1948.

1940–1958
Wartime Boom, Postwar Progress

1940 New director Columbus Iselin offers WHOI expertise for national defense; the first federal contract supports work on marine fouling

1941 WHOI becomes a year-round institution, with research almost wholly devoted to the war effort

1942 *Atlantis* moves to a lake mooring in Louisiana for safekeeping while a variety of large and small vessels work out of the Woods Hole boat basin on WHOI projects

1943-45 Subjects of 40 wartime projects range from antifouling paints and underwater sound to prediction of currents, drift, waves, swell, and smoke screens

1946 Some wartime projects continue and more than 40 WHOI staff travel to Bikini Atoll to conduct before-and-after surveys in the bomb test area

1950 "Iceberg" Smith becomes director and scientists persevere despite postwar adjustments and financial uncertainty caused by wartime inflation

1954 As principal Institution funder, the U.S. Navy completes WHOI's second laboratory building on Water Street to house classified work

1956 Research for the International Geophysical Year helps to sustain WHOI until the Russian Sputniks spark new interest in and funding for U.S. science in 1957

1958 Paul Fye begins his 18-year term as director and *Chain* brings big-ship oceanography to Woods Hole

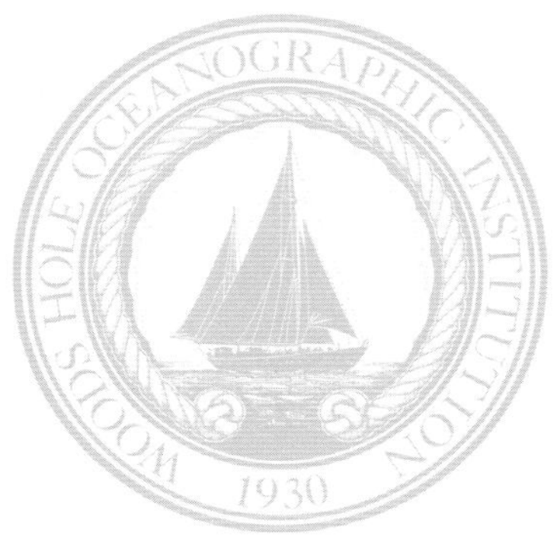

Atlantis pulls away from the WHOI pier for a mid-1940s cruise.

1940

World War II is under way in Europe. As it appears more and more likely that the United States will be drawn into it, Iselin and the WHOI board consider the Institution's potential wartime role. Indeed, war-related work begins on July 1, 1940, under WHOI's first federal contract. The mission is to answer questions posed by the U.S. Navy Bureau of Construction and Repair on the nature of the slime that forms on ship bottoms, its magnitude, organisms that attach to it, and how to prevent this "fouling." Bostwick Ketchum, previously a WHOI summer investigator, is recruited from Long Island University to work with Alfred Redfield on this program at a first-year salary of $3,000.

Iselin keeps in touch with developments in Washington through his Martha's Vineyard neighbor and WHOI trustee Frank Jewett, president of the National Academy of Sciences and chairman of the board for Bell Telephone Laboratories. Concerned that current relationships among government agencies and private advisory bodies are too cumbersome for wartime demands, Jewett and four colleagues[51] persuade President Roosevelt's administration to create an independent mobilization agency called the National Defense Research Committee (NDRC). This committee will organize and administrate both Navy and civilian wartime oceanography projects. Iselin soon finds that he must learn how to deal with the Washington bureaucracy. At one point, he writes to Jewett, "We have found that in writing the Navy one is more apt to get results if each letter is confined to a single subject."[52]

Oceanography and National Defense, 1940

On August 5, 1940, Columbus Iselin addresses a memorandum to the newly formed National Defense Research Committee. His objective is to suggest how "the personnel and equipment of this laboratory can better be utilized for the national defense." He observes that exchange of information between U.S. oceanographers and Navy personnel has been limited, while German oceanographers have been working on "a number of naval problems" over the past five years. Noting that the marine fouling project is well under way, he lists seven other ways that oceanographic knowledge might be useful to the Navy:

- Consulting "on any matter involving geographical, physical, chemical, biological or meteorological considerations."
- Training of naval officers in oceanographic methods and knowledge. "The U.S. Coast Guard has for a number of years had such a policy and it has worked very well."
- Critical analysis of Navy investigations and observations.
- Detailed studies of currents and other characteristics in critical areas of the ocean. "We have good evidence that German oceanographers made detailed studies long before the outbreak of war of the currents in certain critical areas."
- Improvements to and development of instruments.
- Meteorology.
- Use of *Atlantis*. "There may well be questions in distant regions which the Navy would like to look into without attracting attention and in some cases a scientific vessel can be successful in such missions."

Iselin continues his sales mission both in person, visiting his many contacts in Washington, D.C., and through correspondence. On August 22 a letter to Commander R. P. Briscoe, Naval Research Laboratory, says that "it seems likely that a thorough investigation of the transmission of sound in sea water will prove our most useful contribution to your regular research program."

All of these suggestions—except for Navy use of *Atlantis*—prove prophetic.

WHOI employees worked at close quarters during the war. There are few photographs from this period, but this 1946 picture of a group processing hydrographic data is probably typical. Left to right, those shown are Louis Post, Val Worthington, Patricia Brown, Eileen Scharff, Gloria Gallagher, and Doris Lumbert.

In this 1946 photograph, five men work in the machine shop that was set up on the ground floor of the Bigelow Laboratory in 1940. Ralph Bodman is behind the machine at left, and Tex Hoadley and Fred Gaskell are third and fourth from left; the other two men are unidentified.

By the end of 1940, Iselin reports that "a considerable part of our bacteriological and biological effort is directed towards a study of marine fouling organisms," and "the decision of the Trustees at their meeting in August to offer our facilities to the government has led to a contract with the National Defense Research Committee and to a considerable expansion of both equipment and personnel. . . . Our ordinary oceanographic program was at first only slightly interfered with, but since October the entire time of the *Atlantis* and of several of our staff has been required for the new undertakings."[53]

Eight Hundred Miles to Cross the Fence

Two of Maurice Ewing's students at Lehigh University, Allyn Vine and Joe Worzel, had long craved their own machine shop. When they came into Woods Hole aboard *Atlantis* during summer 1940, they persuaded Ewing to offer the owner of the Penzance Garage (located next to the WHOI laboratory) a few hundred dollars for his lathe, drill press, and miller, which were hardly used but rusting from being flooded during the '38 hurricane.

For the trip to Pennsylvania, Worzel and Vine loaded the larger parts of the machinery into a hired dump truck, and they transported the more delicate parts in Ewing's Model A Ford, "Floosey Belle." They had just spent the week before classes cleaning and reassembling the machines when Ewing called them from Woods Hole with an invitation to join WHOI's defense effort. For the trip back to Woods Hole, "We hired another dump truck, loaded up the machines and set them up in a basement at WHOI, thus establishing that institution's first [large] machine shop," Worzel wrote later. "Those three heavy machines had a ride of more than 800 miles to get across the fence at Woods Hole. They were crucial tools for our work during World War II."[54]

The Carnegie Institution closes its Dry Tortugas laboratory and transfers that lab's 70-foot (21-meter) powerboat *Anton Dohrn* to WHOI. With $18,000 worth of Institution-supported reconditioning, the gift provides a heavily built boat with a cruising radius of 2,600 miles (4,160 kilometers) at 8.5 knots, space for a seven-man crew, and moderate fuel costs. During 98 days at sea in 1940, *Anton Dohrn* proves both "able and economical," according to Iselin's account in the annual report. Biologist Gordon Riley has a slightly different take: "The *Anton Dohrn*, built to work in coral reef country, was as funny a little ship as I have ever seen. . . . A double ender with a very shallow draft, she was terrifically heavily timbered. The ribs were so close together that the portholes had to be oval. She had all the sea-keeping qualities of an old fashioned, round bottomed bathtub." [55]

The 70-foot (21-meter) *Anton Dohrn* made 40 cruises from 1940 to 1947 for WHOI investigations from the Gulf of Maine to the coast of New Jersey. Long-time marine superintendent Dick Edwards says it took eight bilge pumps to keep *Anton Dohrn* afloat.

Salaries and appointments approved at the August 15 meeting of the Executive Committee include the following:

- Alfred C. Redfield (Harvard University), associate in marine biology; Norris Rakestraw (Brown University), chemical oceanographer; and Carl-Gustav Rossby (MIT and assistant chief of the U.S. Weather Bureau), associate in physical oceanography—each for five years from September 1, 1940, without salary.
- Henry Stetson (Harvard University), submarine geologist for five years, at a salary of $1,200 per annum.
- Raymond B. Montgomery (New York University), physical oceanographer for three years at a salary not to exceed $2,500 per annum, as fixed by the director.
- Mary Sears (Wellesley College), planktonologist, for three years at a salary of $800.
- The director's salary was set at $4,800 per annum beginning on January 1, 1941.

The BT: Saving Ships and Subs

In summer 1934, Carl Rossby (MIT) took a new invention to sea on *Atlantis*. The "oceanograph" was a large boxlike contraption intended to record continuous tracings of temperature versus depth in the surface layers of the ocean. Rossby was aiming to improve collection of these measurements over the standard procedure: lowering a string of reversing thermometers attached to Nansen bottles that collected water samples. Though the cumbersome instrument vibrated and its linkages were at the mercy of fouling seaweed, it was an important new beginning for oceanographic temperature measurements.

Rossby passed development of the device to his South African student Athelstan Spilhaus, who had a workable instrument by summer 1937. The small torpedo-shaped device contained a temperature sensor and an element to detect changes in water pressure. Because pressure in decibars approximately equals depth in meters, this allowed correlation of depth with temperature. Since atmospheric inventors had already claimed the word *barothermograph*, Spilhaus drew on the Greek root for "depth," *bathos*, and coined the term *bathythermograph*, quickly shortened to "BT."

Spilhaus projected wide application for the BT, from effects of temperature and depth on marine life to the structure of eddies swirling along the edges of the Gulf Stream and other grand ocean currents. However, it was immediately put to use for purposes of national defense.

Athelstan Spilhaus launches an early version of the bathythermograph.

Officers aboard the navy destroyer USS *Semmes,* working in the warm waters off Guantánamo Bay, Cuba, were having difficulty with their experimental sonar (*sound navigation ranging*) gear. It was designed to detect the noise of a submarine's propeller or an echo off a sub's hull from several thousand yards away. Though it had worked perfectly on other occasions, for some reason, in tropical waters its reliability consistently deteriorated in the afternoon. It saw targets that were not there and missed some that were. A call for help to the Woods Hole Oceanographic Institution brought Columbus Iselin and *Atlantis* to investigate the puzzling "afternoon effect" in February 1937, using thermometers and Nansen bottles, and again in August that year, with the BT.

Temperature readings showed that, by early afternoon, the sun had warmed surface water to a depth of about 9 meters (30 feet) until it was 2° to 4°F (1° to 2°C) warmer than the water beneath it. Below the surface layer, the water grew rapidly colder with depth.

> *Knowing that the speed of sound increases with temperature, the scientists realized that signals from the ship's sonar would travel quickly through the warm layer and then slow dramatically when they hit the cooler layer below. They discovered that sound waves passing between layers with different properties underwent refraction, bending away from the region where sound travels faster and toward the region where its speed slows. This bending creates a "shadow zone," allowing any submarine positioned just beneath the dividing line between the warmer and cooler layers of water to become invisible to sonar signals. . . .*
>
> *Columbus Iselin immediately recognized the significance of the acoustic shadow zone and the BT to submarine warfare. A submarine equipped with a BT could use it to determine where the shadow zone lay in relation to the pursuing ship, thus becoming nearly invisible to an enemy sonar. A sub chaser, for its part, could use a BT to opposite effect, adjusting the direction of its sonar to take into account the expected refraction.*[56]

Though the 1937 BT was still cumbersome and slow, needing to be raised and lowered slowly from a stationary ship, and the Navy did not yet consider it of great value, this marked the beginning of a strong alliance between WHOI and the Navy.

Allyn Vine, Maurice Ewing, and Joe Worzel took over BT development in 1940 and soon created a sleeker, more efficient instrument, one that eventually could be deployed and retrieved from a ship moving at fifteen to twenty knots. Vine, Dean Bumpus, and a number of other WHOI people trained ships' officers in BT use in classes held at Woods Hole, aboard ships, and at navy "sound schools." They also developed a novel circular

Thousands of mechanical BTs were launched and retrieved from navy ships and oceanographic research vessels from 1940 to the mid-1950s, when the expendable BT (XBT) became available. Henry Stommel termed the instrument "the murderous mechanical bathythermograph, which despite its propensity to fling its weight about wildly at head-level, never killed anyone."[57] In order to prevent the BT from crashing into the side of the ship and to get as straight a drop as possible, the launch was accomplished by swinging the instrument out and away from the side of the ship on the BT boom before letting it drop, with wire attached, into the ocean. Today's XBT is fired from a launch tube, then trails a thin copper wire that transmits temperature and depth readings to a shipboard recorder until the instrument reaches the end of the wire, breaks it, and falls to the seafloor.

slide rule to speed BT calculations. The next step was to adapt the BT for submarine use (SBT), to provide both sound transmission and ballasting data. During the development phases, two hundred BTs, fifty BT winches, and twenty SBTs were constructed in the WHOI machine shop; later manufacture was transferred to commercial enterprises. After the war, President Harry Truman awarded Columbus Iselin the U.S. Medal of Merit for WHOI's work in developing the bathythermograph. (Iselin said the medal should have gone to Vine.)

The BT slides were sent to WHOI or to a Navy facility, providing the oceanographic community with a valuable database totaling sixty thousand temperature and density (salinity) records by the end of the war. Ship and submarine crews often included notes with the slides. One from USS *Guitarro* said, "The engineering officer is happy to be

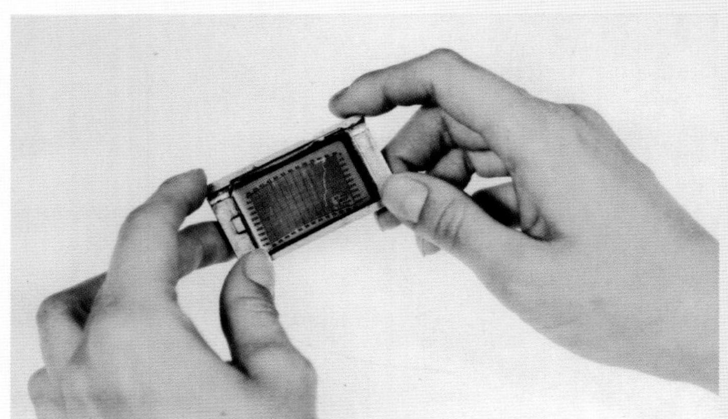

able to forward this card because it means we were able to 'walk away' . . . following a successful attack on a heavy cruiser." Another simply said, "Thank God for Allyn Vine."[59]

1941

During the early 1940s, sixty-three Institution employees leave for active military service. The Executive Committee votes in late 1941 to make up the difference for any whose military pay is less than his or her WHOI salary.

In some cases, it takes considerable negotiation to retain critical staff, including the director. For example, in early January, a Captain Herbster of the Inshore Patrol's Boston headquarters tries to put naval reservists Columbus Iselin and Maurice Ewing on active duty because he needs "someone to develop an effective network of underwater listening devices and radiobuoys along his section of the coast." Iselin appeals to Commander R. P. Briscoe at the Naval Research Laboratory to intervene, adding gracefully, "Of course both Ewing and I will be only too glad to report for active duty, if in that way we could be of greatest use to the Navy." Six months later, this issue is still under discussion, and Iselin hears from Bigelow, "I have just had a letter from Jewett reporting results of his conversation last week in Washington with Admiral Van Keuren and Commander Briscoe. All hands are agreed that you and Ewing can render your greatest service to the Government by sticking to the present job at the W.H.O.I." [60]

Iselin gets involved in clearing Athelstan Spilhaus, a native South African, for war work. In a letter to Spilhaus, then at New York University, on October 21, he writes that he is looking at "a form which the Navy has designed (and it is some job of designing) for cases where it is necessary to employ aliens on secret or confidential contracts. There are nine copies, one of which requires a photograph. However if we all work hard for several days, I believe that the form could be filled in and that we would later be free to discuss the basic problems of our contracts." [61]

With the occupation of Denmark in 1940, stocks of the Copenhagen standard seawater used by oceanographers everywhere soon become alarmingly low. Cooperating with the Coast Guard, WHOI staff members collect and prepare large quantities of water from continental slope water and ensure "a satisfactory

This identification card was issued to Columbus Iselin in 1941.

standard for salinity determinations for the next five years or more," according to the annual report.

In December 1941, Iselin negotiates with the Navy about possible military use of *Atlantis,* and Harvard professor E. Bright Wilson visits Woods Hole to determine whether Institution facilities may be appropriate for his group's investigation of underwater explosives.

Biologists analyze extensive samples taken from Georges Bank waters during a dozen *Atlantis* cruises from 1939 to 1941. They observe that Georges Bank plankton are sharply distinguishable from those in surrounding waters, with *Sagitta elegans* particularly characteristic of Georges Bank water and *Calanus finmarchicus* dominant in the Gulf of Maine current flowing onto the bank. More than fifty years later, scientists conducting the Global Ocean Ecosystems Dynamics program on Georges Bank will mine these data for their studies of the biological characteristics and dynamics of their target species, which include *Calanus.* They will note that the *Atlantis* work offers the only previous depth-stratified data available to them.

The 1941 financial statement from treasurer Lawrason Riggs reports that the government was billed $114,848 for expenses and $21,278 for overhead. "In addition to this a substantial part of our entire income is being used in the interests of the government without reimbursement." And Iselin writes, "Thus by the end of 1941 our regular program of research has been almost entirely superseded by a number of projects of importance to the war effort."

1942

Ninety-six people work at WHOI, and the government is billed $202,719 for operating costs and $40,563 for overhead. Alfred Redfield takes a year's leave of absence from Harvard to become associate director of WHOI; his term will actually last fourteen years. He will also serve as a WHOI trustee from 1936 to 1955; a member of the corporation from 1936 to 1974, and after that as an honorary member; and a member of the Executive Committee from 1943 to 1955.

Atlantis remains a civilian ship (alterations necessary to comply with the larger crew required by Navy regulations prove impractical) and cautiously heads south in February for work in the Caribbean. Increasing enemy submarine activity diverts the ship first to the Gulf of Mexico for two months and then to a mooring in Lake Charles, Louisiana, to ride out the war.

Gordon Riley joins the antifouling project. In an unpublished memoir, he describes it:

> *The Navy Bureau of Ships had developed a number of paints which used copper or mercury or both as anti-fouling agents, embedded in matrices of various sorts. Ideally, the metallic compound needed to be just slightly soluble, releasing the poison at a relatively constant rate that was just sufficient to prevent attachment of fouling organisms and lasting as long as possible. Buck [Iselin] had developed a team of biologists and chemists to work on various aspects of the problem. Testing methods included chemical determinations of the rate of leaching of the metal into seawater over a period of time and submersion of test plates into the ocean to observe their resistance to fouling.*

The Clarke-Bumpus plankton sampler, originally designed by George Clarke and modified by Dean Bumpus, who stands on the *Atlantis* platform during a 1941 cruise, was an important collecting tool during the Georges Bank cruises from 1939 to 1941.

(It also saved shipyards from a hopeless backup of ships needing scraping and painting.)

The project concerns not only the fouling of ship hulls but also of buoys, anchors, chains, amphibious aircraft, and ships' saltwater piping systems. The organisms collected for the antifouling project provide a unique, comprehensive sampling of sessile (attached rather than free-swimming) organisms. To help with testing the antifouling paints, Ketchum gives colleagues paints of various compositions to use on their own boats for the promise of a report, telling them to put one kind of paint on one side and another on the other side. Barnacles, the chief foulers, set only twice a year in Woods Hole waters, so a lab is set up in Miami, where barnacles set year round. The project's boathouse headquarters is the forerunner of the University of Miami's marine laboratory.

Al Woodcock, left, and a colleague run a smoke generator during World War II studies of potential use of smoke screens during beach landings for troops.

The wartime security gate was a busy place. Columbus Iselin is at center, security guard Harry Handy is at right, and Kenneth McCasland is between them. The two men at left are unidentified.

1943–1945

Because most wartime work is classified, the slim thirty-five-page volume published as the collective annual report for these years belies extensive WHOI defense research. The crowded laboratories hum through long days and into the weekends. The "white house" goes up behind the laboratory to house administrative offices. At least forty projects are under way, some of them subsets of others, according to an unpublished outline Iselin writes later.[64] For easy reference, the projects receive numbers and are referred to as "Navy One, Navy Three, Navy Seven." For military purposes, Navy Two is referred to as "sound transmission in seawater;" for civilian purposes, it is known as "temperature distribution in the surface layer of the North Atlantic."

Allyn Vine contributes to several projects. Writing in a 1971 issue of the Institution's *Oceanus* magazine devoted to Columbus Iselin, he provides a short summary of wartime activities and research subjects:

- performance of antifouling paints (see pages 61–62)
- transmission of sound in the sea
- development and application of underwater explosives
- underwater photography
- current and drift predictions for downed aviators on life rafts
- prediction of wind shear and inversion layers on smoke screens
- burying and countermining in mine warfare
- wake suppression for landing craft
- wave and swell predictions for amphibious landings
- submarine diving problems

A temporary structure behind the laboratory dubbed the "white house" accommodated administrative offices from 1944 until 1969, when a new, larger pier was constructed. Beginning in 1943, the Penzance Garage, to the right of the white house, housed welding, brazing, pipe cutting, and blacksmith and tinsmith works. The parking lot on the street side of the garage was displaced in 1958 when construction began for the Laboratory of Oceanography (now the Smith Laboratory), built by the Navy especially for classified work. The new structure incorporated part of the garage.

Four men work in a Navy Seven (explosives project) laboratory on the top floor of the Bigelow building.

- convoy routing

- discouragement of sharks

- writing of manuals to instruct Navy people of all ranks how knowledge of the ocean might improve their situation

Navy Seven moves its underwater explosives investigations from Harvard to WHOI when Cambridge neighbors complain about the noise. Staff members of this project include Paul Fye, who will later serve the Institution as a forward-thinking director for eighteen years; several men who will play important roles as WHOI board members; and Betty Bunce, who will become the first woman to serve as chief scientist at sea and as acting department chair ashore. This group takes over the laboratory's top floor to develop better chemical explosives and determine how to maximize their effects both in water and in air. The Forbes family allows use of its Nonamessett Island across Woods Hole Harbor from the laboratory, and it is here that Navy Seven stores, mixes, and casts explosives. WHOI acquires *Reliance* and other boats to transport people and cargo to and from the island and to carry out experiments. Train cars filled with explosives arrive in Woods Hole, test explosives rattle (and break) windows, and the group earns the moniker "boom boys." They develop instruments and techniques to observe shock waves from a wide range of charge sizes, examine the bubbles that explosions produce, study damage to structures

[Navy Seven] went to bigger and bigger charges and fancier chemical mixes. We had to move the underwater operations out into Vineyard Sound and the air measurements to isolated parts of the Cape. . . . [The group] had a young physical chemist named [Don] Hornig. As far as I was concerned he mysteriously disappeared from the payroll. He devised the way to explode the first atomic bomb [what height to fire it at], based on measurements that we had made off our wharf. . . .

—Columbus Iselin, unpublished personal history, about 1965[65]

(including submarine hulls), and compare the effectiveness of various explosives. Their work includes testing full-scale weapons and fusing devices, studying countermining, and determining optimum height of detonation in air. The latter contributes to the effectiveness of the bombs dropped on Japan.

Iselin writes of Navy Seven that "from the standpoint of oceanography this work has been of value because much experience has been gained in carrying out elaborate and precise experiments at sea. The use of complicated electronic instruments on shipboard is a relatively new problem, but in all probability it will be a continuing one in oceanography."

Paul Fye ponders what to do with an uncooperative camera rig during a 1945 Navy Seven *Atlantis* cruise to study underwater explosives.

Woods Hole village is forever limited in housing for WHOI employees. Gordon Riley describes one solution during the 1940s: "[Hank Stommel] and several other young bachelors rented a house that had formerly been an Episcopal rectory and that was still called the rectory, although I dare say it was gayer and

The fleet expanded to as many as 14 large and small vessels during the war. This photo from summer 1945 shows the 142-foot (43-meter) *Atlantis* at upper left and the 70-foot (21-meter) *Anton Dohrn* at upper right. *Dot II* is below *Atlantis* with *Lobster* tied to it. *Esk*, aft of *Dot II*, ferried Columbus Iselin and others on their daily commute from Martha's Vineyard, served as a tug for *Atlantis*, and also engaged in research activities. Upward from the middle bottom of the photo, the lineup is the *Reliance*, *Asterias*, *Mytilus*, and *Clare*. The rest of the fleet comprised *Everette*, *Senior*, *Reliance*, and *Physalia*. The addition of government-financed boats during the war kept a crew of ship carpenters, electricians, and mechanics hopping, under the direction of John Churchill.

noisier than it had ever been in its earlier days. . . . I remember a Halloween party . . . where they had gone to considerable effort to create fearful and way out décor right from the front entrance through the whole downstairs area, and all dimly lighted, providing lurking places for hobgoblins in every corner. Hank and the other young residents of the rectory typically were about as energetic and imaginative in their fun as they were in their science." [66]

Atlantis returned to Woods Hole from its safe harbor in Lake Charles, Louisiana, in July 1944. On September 14 the ship was moved to the Fisheries dock because of the impending hurricane. Captain Lambert Knight and four other men stayed aboard. When the fastening lines parted, Knight managed to steer the ship toward a place he knew to be mud-bottomed. [67] After the storm, *Atlantis* and USS *Saluda* rested on Ram Island flats in the upper reaches of Woods Hole Great Harbor near the causeway to Penzance Point. The smaller *Saluda* was soon floated, but drawing 17 feet (5 meters), *Atlantis* was, for all practical purposes, high and dry in 6 to 7 feet (2 meters) of water. There it rested on its bilges for three weeks until a salvage operation could be arranged. There were no major damages to *Atlantis,* but salvage costs exceeded $17,000.

Germany surrenders on May 7, 1945, and Japan on September 2. By September 15, WHOI's special security status is cancelled, and postwar readjustment begins for an institution where everything has changed, beginning with a staff increase from less than 100 before the war to more than 300 in 1945.

1946 For the time being, the accelerated wartime pace continues at WHOI. A new, postwar contract with the Bureau of Ships supports work in underwater sound (including extensive temperature and salinity measurements in the Atlantic), submarine diving and ballasting, ship fouling, development of instruments related to undersea warfare, waves, surf, meteorology, and consultation services for the U.S. Navy Radio and Sound Laboratory. In addition, more than forty staff (of about three hundred) participate in Operation Crossroads, U.S. atomic bomb tests at Bikini Atoll. As the military prepares to evaluate the destructive powers of nuclear weapons by detonating two bombs—one under water and one in air—at the Pacific island,

> *Now that the war is over and the security regulations have been relaxed, it has become possible to present a general picture of the part played by this laboratory in the war effort. It should be emphasized at the outset that we at Woods Hole were by no means the only group in this country working on the military applications of oceanography, nor should we be given credit for more than a minor share of the successes. First under the leadership of the National Defense Research Committee and more recently under the direction of the Bureau of Ships and the Hydrographic Office all laboratories concerned with problems involving the sea have worked in closest collaboration. Each group conducted investigations for which their facilities and experience were best adapted, in most cases only part of a given project being carried out at any one laboratory. This "production line" system was on the whole very effective, the final assembly usually taking place within the Navy. In this way the limitations of a civilian organization in war time were largely overcome.*
>
> —Columbus Iselin, 1943–45 WHOI annual report

> *Any portrait of Iselin at Woods Hole without mentioning John Churchill would be deficient. John was a classmate and shipmate of long standing, and in the "Log of the Schooner Chance" was called "the hard-boiled engineer." By profession he was an architect, but his heart was on the water, and his mode of thinking and action were obviously in line with Columbus'. As a result, Columbus had an executive who also believed in shoving great responsibility on individual scientists and workers, but who was quite capable of making major decisions, skippering the ships and arbitrating disputes in a way that was more likely to leave the participants laughing than mad.*
>
> —Allyn Vine, Oceanus magazine, June 1971

From left, Arnold Clarke, Ruthann Wyrick, Charlotte Vail, and Bobby Atwood attach thermometers to Nansen bottles for shipping to the Pacific for surveys around Bikini Atoll prior to the atomic bomb tests there.

the U.S. ocean science community successfully lobbies for a before-and-after Bikini research program focusing on ocean currents, waves, instrument development, and the effect of both radiation and the force of the blast on marine life. In a March 1946 letter to Mary Sears, still at the Hydrographic Office (see feature on pages 86–87), Iselin characterizes this opportunity for ocean scientists as "the biggest thing scientifically that will ever happen during our productive years" [68]—presenting the chance to use many ships and expensive facilities to develop a comprehensive understanding of the physical, biological, geological, and chemical characteristics of a broad oceanic region. The Hydrographic Office and WHOI, with Gordon Riley as senior scientist, take the lead in physical oceanography—one participant borrows "every single reversing thermometer in the country" for the survey. [69]

I had a team of two dozen to take to Bikini to study the current systems, both in the lagoon and outside, in the event that a slug of radioactive water got out of the lagoon, and also . . . for the underwater explosion, how would the radionuclides move.

—Unattributed comment at 1991 WHOI history colloquy

The war has effected a permanent change, and there is worry about the Institution's future. In the 1946 annual report, Iselin laments,

> It has become evident that postwar costs of oceanographic field work have more than doubled as compared with the period just before the war. This is perhaps the most serious problem facing the Institution today. The costs of operating our own vessels have become so high that without Government subsidy we could not hope to undertake offshore observations. For the time being, at any rate, the government is able to support science at private laboratories on an unprecedented scale, but should we have to return to operating solely on income from endowment, we could no longer afford more than a small program of field observations in local waters. . . . Facing the prospects of a greatly reduced field program within the next year or two, we have deliberately exerted every effort to keep our vessels busy accumulating observations.

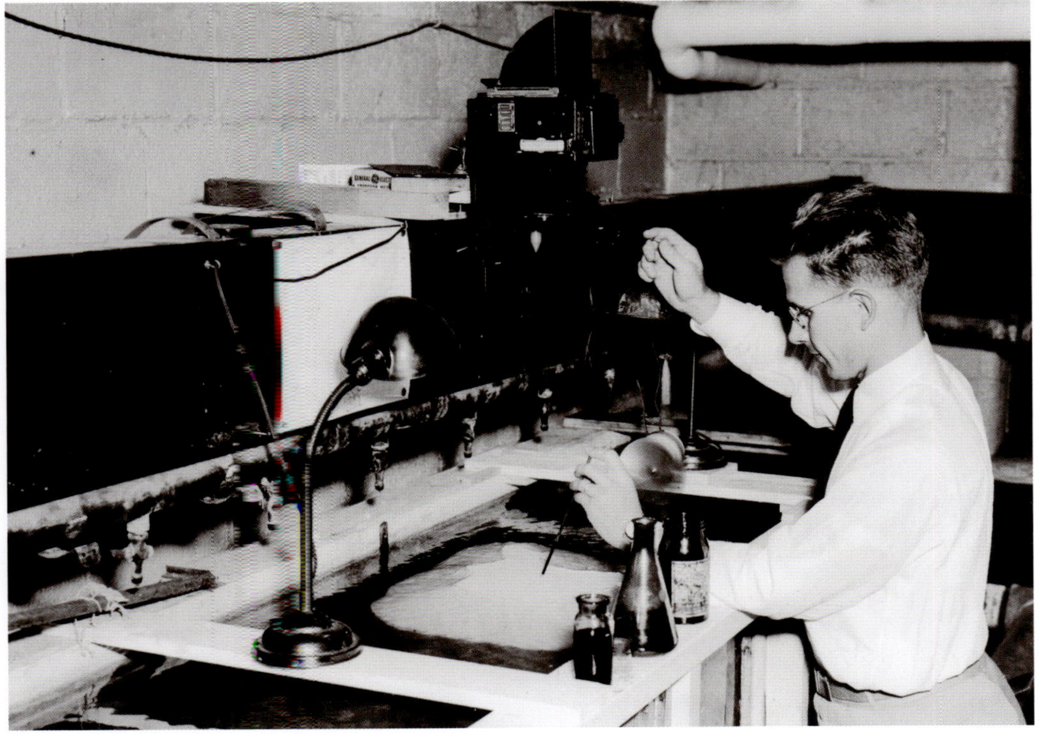

Bill von Arx was part of the Bikini Atoll group studying circulation inside the island's lagoon. Realizing that the bomb test was scheduled for a period of light winds, while the prior studies had been done under strong trade wind conditions, he returned to Woods Hole to make this model of Rongelap Lagoon using a flume left from war work on sand movement around mines. With it he correctly predicted where in the lagoon the radioactive water would concentrate.[40]

Fortunately, work with oceanographers during the war has convinced the Navy that many of its operations are intimately dependent upon the environment in which its ships operate—that is, upon a basic understanding of oceanography. Iselin carefully monitors the Navy's rocky transition from wartime mobilization to peacetime reorganization and then Cold War demands. The Hydrographic Office has a new Division of Oceanography, and August 1946 legislation creates the Office of Naval Research (ONR). In its early years, ONR provides "block funding," large contracts to major oceanographic institutions that rely on the principal investigator's experience and seasoned judgment to support productive projects. Iselin maintains WHOI engagement with Navy affairs as a member of the newly formed Committee on Undersea Warfare. He is, however, always mindful of preserving oceanography's academic integrity, and his correspondence in the postwar period returns "frequently in both professional and casual letters to the need for careful disciplinary definition and the construction of a culture resting upon university-based educational programs directed by experienced oceanographers."[71] Along with other leaders in the field, he worries that definition of the still-infant science of oceanography can only be achieved in "a few centers of oceanographic research where the subject as a whole can be developed in an orderly manner and where applications are not the main thing. Must attract first rate minds to the subject."[72]

Spotting Talent

Columbus Iselin made up for the lack of qualified oceanographers by nurturing those in whom he saw scientific potential. In a short biography of Iselin, Henry Stommel wrote:

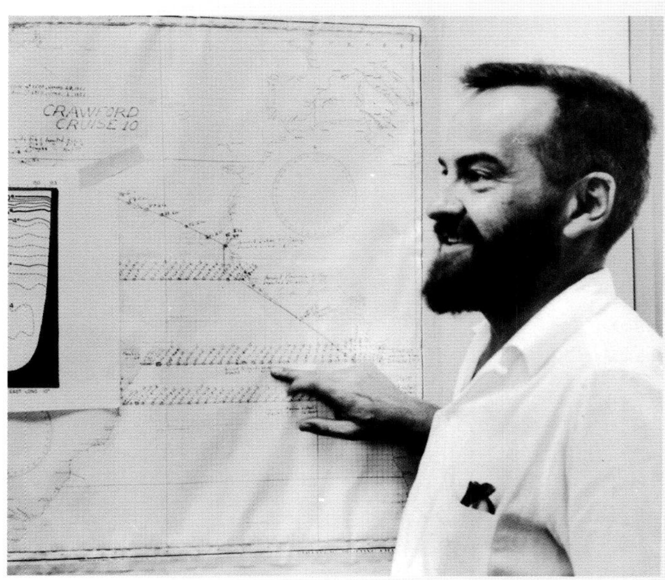

Fritz Fuglister discusses a temperature profile chart.

The development of the B.T. was important for oceanography after the war because it meant that detailed data could be obtained from a ship under way. Of perhaps even more importance was the availability after the war of Loran, which improved navigation to the point where detailed surveys could be made meaningfully. The emphasis in Gulf Stream research shifted from the intractable problem of long-period variability of the Gulf Stream to studies of the detailed structure of its front. Frederick Fuglister was encouraged by Iselin to conduct this research. Fritz's training as an artist had accustomed him to visualizing form, and soon he was bringing back a series of wonderful surveys that showed us for the first time the meander and eddy structure of the Stream.[73]

Artist-turned-oceanographer Fuglister eventually served as chair of the Physical Oceanography Department, as did Val Worthington, who had spent three rather leisurely years at Princeton doing Latin studies before joining the WHOI staff in 1941. (Fuglister,

Worthington, and Stommel poked fun at their lack of science credentials by creating the Society for Subprofessional Oceanographers [SOSO] in 1961 after reading an ONR newsletter announcement about a new European laboratory staffed by eight Ph.D.s and fifteen "subprofessionals." Letterhead created for the society by Fuglister's son named Stommel president, perhaps because he did at the time hold an undergraduate degree in astronomy; Fuglister, vice president; and Worthington, ambassador to the Court of St. James. Though the society didn't hold meetings or keep records, Fuglister staged a coup and took over the presidency in 1964 while Stommel was in California to receive the Sverdrup Medal of the American Meteorological Society.[74]

Another Iselin protégé was Al Woodcock, who joined the original crew of *Atlantis* while working in an apple orchard as part of a program at the Massachusetts Agricultural College. "He has been scientifically the most productive of my students and close friends," Iselin wrote.[75] "From the outset it was obvious that Woodcock was much more than a young sailor. I encouraged him to keep simple records of sea conditions and weather. . . . he became the continuing laboratory technician on the ship and took great interest in the accurate determination of the salinity of the many water samples . . . being collected. Without the use of these careful measurements neither one of my two major papers would have amounted to much. While at sea he also kept a careful log of the flight tactics of sea birds, of the distribution of Sargasso weed and of the sailing characteristics of *Physalia* [the Portuguese man-of-war, whose float has aerodynamic properties]. These data led to the publication of his first three important papers, which are fundamental to the understanding of air-sea interaction." Woodcock later became a professor at the University of Hawaii.

Val Worthington rigs a Nansen bottle for a hydrocast aboard *Atlantis*.

Al Woodcock reads a wind gauge.

After fifteen years of hard work, *Atlantis*'s engine needs replacement. The Navy provides one from surplus, and WHOI reserve funds support rearrangement of the engine room, including a new forced hot-water heating system. The ship emerges from two months at Electric Boat in Groton, Connecticut, "far better equipped with propelling and generating machinery than when she was new," Iselin writes in the annual report.

The Institution continues to operate three government-owned vessels, *Mentor* (127 feet; 38 meters), *Reliance* (76 feet; 23 meters), and *Claire* (a small launch). Iselin writes,

> *In assuming the responsibility for . . .* Mentor, *we have taken a new step which constitutes a rather crucial experiment for oceanography. More and more the facilities and instrumentation needed in the earth sciences are owned by the Government. Does this mean that the private laboratories will gradually give way to the Government laboratories because they cannot afford the elaborate new equipment? I believe that so specialized a vessel as the* Mentor *can be more efficiently operated by a civilian crew than by a navy crew and events seem to indicate that the Navy has sufficient confidence in the Institution to give this plan a try. Should we fail in this responsibility, then it seems likely that the Government will of necessity seek other means of developing the earth sciences. Surely the Government as the only agency now capable financially, must encourage, in one way or another, continued development of the earth sciences. Unless man continues to broaden his knowledge of the earth, including the oceans, he will not be able to develop fully the resources of his environment.*

WHOI biologists return to their prewar emphasis on the productivity of the sea, focusing on coastal waters, and they begin marine pollution studies. Iselin notes in the annual report that a book published in 1945 on this subject illustrates "what a substantial contribution to the subject as a whole has come from Woods Hole." Observations in trade-wind areas signal the beginning of what will become a strong meteorological group (see "Earth's Greatest Interface" on pages 140–41). Ewing's former student Brackett Hersey joins the staff to head a group that continues underwater sound research, receiving reflections from "considerable distances" below the bottom and investigating the "deep scattering layer" that plagued submarine commanders during the war. These reflections from marine animals migrating in response to changing light through the day will be the subject of WHOI research for many years.

Laboratory space and housing continue to be in short supply. WHOI rents MBL laboratories in winter and the garage year round and also purchases the Hall property across Water Street from the garage.

While working long hours on wartime and postwar projects, WHOI staff also performed in 1945 and 1946 stage shows called the Woods Hole Follies and, in 1961, in a follies revival. Joe Worzel, left, and George Woolard share the stage in the 1946 show.

1947

Work at Bikini Atoll continues, and two major *Atlantis* cruises in 1947 respond to both oceanographers' desire for blue-water research and Navy interest in seafloor topography and underwater sound transmission. Cruise 150 in July and August is the first since 1940 for chief scientist Maurice Ewing, who moved to Columbia University in 1946. It yields sediment cores from Henry Stetson's coring device, echosounder profiles of both the abyssal plain and the mid-ocean ridge's spectacular peaks, a quarter-ton of volcanic rock from the dredge and—using a ton of TNT—extensive seismic records, including the first made while underway rather than on station.

Then, in mid-December, on Cruise 151, *Atlantis* returns for the first time to European waters. Funded by the Hydro-

Maurice Ewing's Columbia University student Frank Press studies seismic records on *Atlantis*'s deck during a 1947 cruise. Press later played important roles in ocean sciences as head of the MIT Department of Earth and Planetary Sciences, where he facilitated the creation of the MIT/WHOI Joint Program; as president of the National Academy of Sciences; and as science advisor to President Jimmy Carter. Press was the keynote speaker at the Institution's fiftieth-anniversary dinner dance, where he delivered a congratulatory message from the president.

Valletta, Malta, was among the ports *Atlantis* visited during "the Med cruise," a 1948 cruise to the Mediterranean Sea.

In 1948, Nat Corwin, left, and Dean Bumpus plot bottom contours of the Aegean Sea during "the Med cruise."

Dave Owen launches his camera rig during "the Med cruise." The mid-1940s brought significant advances in deep-sea photography and other technology.

graphic Office, the primary purpose of the six-month "Med Cruise" is preparation of bathymetric charts of the Aegean Sea for use by U.S. submariners in the event of a Soviet naval threat from the Black Sea. The cover story for the mapping presents less ominous-sounding science objectives such as physical and chemical sampling, plankton hauls, seismic reflection, camera stations, wave measurements, and many other oceanographic observations in a little-known body of water.

WHOI purchases the eleven-acre Fay estate in Woods Hole and initiates plans for housing, recreation, and a mess in its four residential buildings and two barns.

In the lean postwar years, Iselin encourages the study of coastal waters, which have largely been neglected by oceanographers. These studies lead to the first adequate theory on estuarine circulation. It will prove to be of great practical importance for work in pollution, coastal engineering, and fishery problems. Mary Sears resumes plankton studies, and there are investigations of red tide in Florida, the productivity of waters on coastal banks and in partially enclosed bodies of salt water, and experimental studies of clam culture on seven acres in a Barnstable Harbor salt marsh.

Balanus, purchased in 1946 from military surplus, makes twenty-six cruises by 1950. Of a 1947 *Balanus* cruise in nearby coastal waters, Gordon Riley later wrote:

> *The ship was the* Balanus, *a tubby little vessel with seakeeping qualities little better than those of the* Anton Dohrn *[see page 55]. I don't know who christened her with the name of a barnacle, but it was apt. She never should have come unstuck from the pier. That cruise was a disaster from start to finish and without doubt the most miserable one I have ever been on. The weather really was no worse than average for that time of the year —a five to ten foot swell running most of the time. The* Balanus *was what did us in*

—a quick, hard roll that made deck work miserable
and lab manipulations virtually impossible. We took
turns on the pulpit during bottle stations. And everyone
got dunked to the waist sooner or later. I particularly
remember the last station before we finally gave up and
headed home. It was Buck [Ketchum's] turn on the
pulpit. He had only one set of dry clothes left, and he
wasn't going to get them soaked. He manned the station wearing only underpants and rubber boots while
a spring snow flurry swirled around his bare torso.[76]

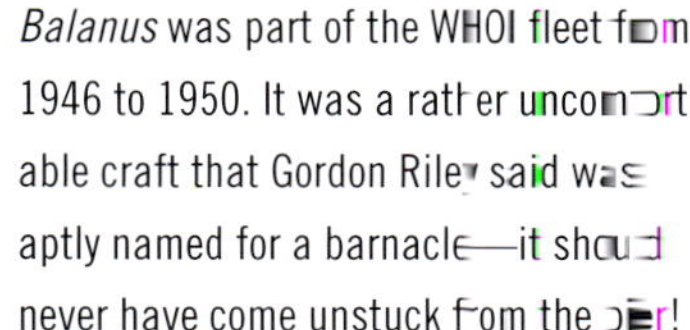

Balanus was part of the WHOI fleet from
1946 to 1950. It was a rather uncomfort
able craft that Gordon Riley said was
aptly named for a barnacle—it should
never have come unstuck from the pier!

From the late 1940s to the 1960s, a team
led by Alfred Redfield investigated clam
farming and the biology of softshell clams
in a large harbor area in the town of Bar-
stable, on the north side of Cape Cod. In
the photograph, Redfield plows the marsh
to help seed clams dig in. This hard work
turned out to be unnecessary as the clams
managed to dig themselves in nicely. The
group learned a great deal about clam
predators and that clam farming should
not be undertaken lightly. Eventually they
expanded their work to study environmental influences on the reproductive cycle of
a variety of bottom-dwelling organisms.

Geophysicists capitalize on wartime advances in acoustics to study the structure of the earth under the ocean basins. Fred Phleger and Henry Stetson, supported by the Geological Society of America, are about halfway through what Iselin calls "by far the biggest job we have yet tackled outside of government contract work," analysis of five hundred cores collected by *Atlantis* in 1945. An ONR grant establishes the Marine Foraminifera Laboratory for Phleger to study the ecology of modern forams.[77]

Iselin notes that salaries at WHOI have not kept pace with the increased cost of living and that, unless this can be remedied, WHOI may have to "resume being a laboratory for summer visitors from the universities with a few full time graduate student assistants. This course would certainly simplify the director's job, but I doubt that knowledge concerning the ocean could be advanced at anything approaching the present pace. . . . [T]he salary difficulties . . . boil down to whether or not this sort of a laboratory can continue to count on government subsidy in peace time. Perhaps these matters will become clearer when the new Congress meets next winter."[78]

Institution staff members serve as advisers to new marine laboratories at the University of North Carolina, on Chesapeake Bay, and in Hawaii.

WHOI's first venture into public relations begins with hiring Jan Hahn to write and facilitate articles for national publications. Beginning in 1952, when the Associates program is initiated, he will also provide excellent photographic and written chronicles in *Oceanus* magazine "to acquaint the many friends of the Woods Hole Oceanographic Institution with the research in progress."[79]

1948

Following outfitting over the winter that includes new frozen- and fresh-food storage and a thousand-gallon fuel storage tank, *Atlantis* sails east on June 15 for a voyage that includes 21,000 miles (33,600 kilometers) of continuous fathometer records, 2,750 bathythermograph lowerings, and 360 plankton tows. The winch cable parts off Dakar because of poor spooling, and *Atlantis* loses its propeller about 500 miles (800 kilometers) off Barbados but manages to return on schedule to New London under sail without incident—"an excellent example," Iselin writes, "of the need for sail in oceanographic vessels." The

On at least three or four
occasions, Hank Stommel and I cooked
up ideas that we thought were good and
worth propagating, and that we ourselves weren't authorized to spend any
money on. One of them was that Buck
Ketchum had got us interested in estuary dynamics. And Hank and I had
done a little theoretical paper on an ide-
alized situation, then decided that we'd
like to have a flume in which we could
study the flow of fresh water into salt,
and the dynamics of the salt wedge. . . .
We went to Columbus and said we
thought this was something that the
institution ought to explore, and this would connect with Buck Ketchum's interest
in estuaries on the biological end. And his answer was, "sure, go ahead." And we
hired Harlow Farmer, who had gotten a degree in civil engineering at MIT, and he
and Ted Spencer built a flume for these studies.

—Arnold Arons, WHOI history colloquy, 1991

This flume was located in a small building next to the white house, behind the
Bigelow Laboratory, for studies of phenomena that included the flow of fresh-
water into saltwater. From left, Arnold
Arons, Ray Montgomery, Henry Stommel,
and Harlow Farmer observe the flow.

new propeller, better suited to the still-new main engine, increases cruising
speed under power by nearly two knots. The $10,000 rebuilding of the winch
seems especially urgent. "Because of the very long cores which can now be
secured as a routine by the piston-type tubes the time scale of the sedimentary
studies has been greatly extended. Today a 30 ft. core is as easily obtained as a 6
ft. core by the former method. Since the rate of sedimentation averages perhaps
1 cm per 1000 years, it seems likely that through study of the new cores much of
the geologically recent history of the earth can rather quickly be worked out."[80]

The 98-foot (30-meter) sailing vessel *Caryn*, purchased in 1947, is outfitted for
oceanographic work and completes two cruises in the waters around Bermuda.
One cruise collects marine fauna for the Chicago Natural History Museum
and the Bermuda Biological Station for Research. The other supports Brackett Hersey and Allyn Vine as they refine a discovery vital to Navy interests:

shallow, low-frequency sound returns to the surface at predictable intervals, known as convergence zones, that allow transmission and detection ranging close to the surface.

Faced with an uncertain financial future and the possibility that *Atlantis* and *Caryn* may not be supportable (during this time period, promises of Navy support for their work are not always fulfilled), the Institution continues to emphasize fieldwork whose data may occupy researchers when ships are less available to collect it. By this time, scientists understand the Gulf Stream in considerable detail from Cape Hatteras to the longitude of Halifax but are anxious to expand their knowledge of the current system as a whole. "North and east of the Grand Banks the topography of the bottom probably plays an important role in determining the paths of the various branches into which the Gulf Stream seems to break up. Unfortunately, we still even lack a reliable chart of the bottom topography as a starting point for an understanding of the current system as a whole." What they are able to surmise is that "the permanent currents in the ocean are mainly the result of the distribution of temperature rather than its cause, in other words, that we are dealing with a huge heat engine and that only part of the energy comes from the winds.

Built in 1927 (possibly for smuggling opium) and previously known as *Black Swan, Santa,* and *Marie, Caryn* served as a platform for a variety of investigations during the 110 cruises it made for WHOI from 1948 to 1958. Rechristened *Black Swan* by a new owner, the vessel burned in the West Indies on New Year's Eve 1974.

This is the very opposite to the approach to the circulation problem made popular by the Norwegian school of oceanographers and brought to this country by Rossby and Bjerknes."[81]

Costs for repairing and modernizing *Atlantis* have averaged about $25,000 per year since the war, and Iselin speculates that "this may continue for several years until almost all but the hull has been replaced."[82]

A postwar exodus from WHOI continues in 1948, with scientific staff taking appointments at Navy labs and universities from Cornell to the University of California, but usually retaining a summer connection to Woods Hole. There are 264 people working for the Institution as of July 15, 1949, including those on fellowships (twelve) and visiting investigators (thirteen).

A long coring device goes over the side of *Atlantis*. Extensive coring of Atlantic Ocean sediments began in the mid-1940s.

1949   With WHOI's future still unclear, the impending Korean War brings welcome new support to oceanography—"and this money we are almost entirely free to spend as we think best," Iselin says in the unpublished 1949 annual report. He predicts that research on the borderline of oceanography and meteorology will be urgently needed—"I am hopeful that oceanographers can make a rather substantial contribution to the science of meteorology as well as to the operating needs of the navy in the event of another large scale war at sea."[83]

Atlantis spends 235 days at sea and *Caryn* about half that. *Atlantis*'s mainmast is shortened by 28 feet (8.5 meters) from its original 138-foot (42-meter) height because of wood deterioration, and the trawl winch is rebuilt. (Ewing now happily reports securing a 57-foot [17-meter] core in an hour and a half.) Adrian Lane continues as captain.

After reading a Russian paper (written in French) on beluga whales, Bill Schevill and his wife, mammalogist Barbara Lawrence, travel to Quebec to find the animals. "We were able to make these rather inferior records of this wonderful racket," Schevill said. "All we had to record with was a dictating machine . . . but it was ever so much better than trying to describe the sound."[85] This recording, made by lowering a hydrophone from the shore of the Saguenay River, is the earliest one made by WHOI researchers in a library that, by 2005, will grow to tens of thousands of marine-mammal sound records, some contributed by researchers elsewhere.

In an article entitled "North Atlantic Oceanography" for the August 1949 issue of ONR's *Research Reviews*, Iselin writes that *Atlantis*, the only U.S. nongovernment vessel that cruises wide areas of the ocean, is a good platform for perfecting oceanographic instruments. He describes a typical cruise as lasting about two months and covering about 6,000 miles:

> *The recording Fathometer is turned on when the vessel reaches the edge of the continental shelf. Until the end of the voyage, this precious tape must be carefully correlated with the navigational record. . . . Every half hour a bathythermograph lowering is made and every half hour a hydrophone is streamed to obtain the arrival times of the echoes of a small explosive charge tossed over the stern. In this way, the thickness of the unconsolidated sediments on the ocean floor is obtained. . . . About twice a day, the* Atlantis *is hove to for an hour and a half or more in order to obtain a series of deep water samples and to observe the temperature distribution in the layers below the limit of the bathythermograph. . . . The development of the piston-type coring tube, which can bring up sections of the bottom sediments thirty or more feet in length, is a strong inducement to stop at least once each day. . . . In good weather it is possible to obtain deep water samples and temperatures with the hydrographic winch at the same time that the heavy coring tube is being lowered on the trawl winch. Although there is some chance of fouling the two long cables, the time saved often justifies this risk. Sometimes a third lowering on still another winch is risked in order to obtain quantitative plankton hauls or other biological observations.*

All this, he writes, costs about $600 a day. (In 2004, a day of *Atlantis* time—without *Alvin*—will cost nearly $25,000.)

1950

Operation Cabot (originally called Operation Dream) is a 1950 highlight at sea. Using three U.S. government ships and one from Canada, plus

Harold Backus (born Harold Backhouse in the north of England) was a WHOI fixture for many years. Iselin hired him in Woods Hole as chief engineer for *Atlantis* while the ship was still under construction. Backus operated the new Institution's facilities ashore until it was time to go to Copenhagen to prepare *Atlantis* for its first voyage. He stayed with the ship for twenty years, after which he again worked ashore. Iselin said, "He could repair anything.... He was always cheerful and willing, and the best of shipmates under adverse conditions." In addition, Harold's daughter, Jeanne, was Iselin's "right hand" for many years.[84]

Bill von Arx and Henry Stommel developed a new and powerful method of observing surface currents from a ship under way. It depended on the electric potentials induced by the movement of the water through the earth's magnetic field. In this 1948 Med cruise photo, the men in the foreground (from left, James Hollis, Gene Krance, Dean Bumpus, and Hans Cook) watch the readout from the geoelectromagnetokinetograph, known as the GEK (pronounced "geek"). Nat Corwin launches a mechanical Watson current meter aft, with Dave Owen photographing him.

Atlantis and *Caryn*, the Navy's Hydrographic Office mounts a full-scale attack on the Gulf Stream. From June 7 to 22, the six ships travel from Cape Hatteras to the Grand Banks, composing a comprehensive picture of the stream's changing position, surface velocities, meanders, spin-off eddies, and interchange with the atmosphere. WHOI analyzes the data under the leadership of Fritz Fuglister. (See "The Atlantic's Mighty Meandering River" on pages 45–47.)

On July 1, Edward H. "Iceberg" Smith succeeds Iselin as director. He is the first director to occupy Meteor House on the Fay property, and it has been winterized and redecorated for his arrival.

"Iceberg" Smith, Third Director

When Rear Admiral Edward Hanson "Iceberg" Smith became director of WHOI in 1950, he brought a background in both science and seafaring. Descended from a long line of whalers, he was born on Martha's Vineyard with the sea in his blood. He graduated from what became the Coast Guard Academy in 1914 and served more than forty years in the Coast Guard, much of that time with various assignments and commands in the International Ice Patrol (hence the nickname "Iceberg"). Henry Bigelow was the patrol's scientific adviser, and Smith worked with him at the end of each ice season (which lasts from March to July) analyzing data and running experiments on ice samples. He was awarded a Harvard master's degree in oceanography in 1924.

He then traveled to Norway and Britain to study with physical oceanographers and meteorologists, and in 1928 he led the Coast Guard cutter *Marion* expedition to the Labrador Sea, Davis Strait, and Baffin Bay. His thesis for a Harvard Ph.D., awarded

Edward H. Smith, right, WHOI director from 1950 to 1956, greets *Crawford* master David Casiles upon the ship's arrival in Woods Hole in 1956. Smith spent forty years in the U.S. Coast Guard and was instrumental in the transfer of *Crawford,* a former coast guard cutter, to WHOI.

in 1934, concerned a practical method for measuring ocean currents. As one of only two Americans aboard, Smith served as navigator for the German *Graf Zeppelin*'s longest nonstop flight, six days and 8,000 miles (12,800 kilometers) over the Arctic Circle in 1931.

He received the World War I Victory Medal and the Distinguished Service Medal for World War II service as commander of the Greenland Patrol and commander of a task force in the Atlantic fleet. The latter citation mentions naval base construction in Greenland and the Arctic, "vital weather, patrol and escort services," and "splendid diplomacy, sound judgment and intelligent planning" as well as "superior tactical knowledge and steadfast devotion to duty." [87]

Smith was commander, Third Coast Guard District, New York, and captain-of-the-port of New York when he retired in June 1950 to become WHOI director on July 1. His six years as director was a time of significant growth for oceanography in general and WHOI in particular, including a staff increase from about 250 to 350, the beginnings of more formal administrative and financial organization, construction of the Laboratory of Oceanography (later renamed Smith Laboratory), and acquisition of R/V *Crawford.*

Our wartime experiences had driven home the great value of giving the individual scientist freedom to choose his own problem and his own methods of approach. The policies and practices we developed for support of the research of scientists in universities and nongovernmental laboratories have since been improved and refined by the National Science Foundation, the National Institutes of Health, and other Federal research-supporting agencies, but their essentials have not been much changed.[88]

—Roger Revelle, *Oceans*, 1969

As early as July 1945, Vannevar Bush, then director of the Office of Scientific Research and Development, had called for a civilian science foundation.[89] Since then, legislation regarding some form of national science foundation has been introduced annually and finally passes in 1950. The National Science Foundation (NSF) begins operation the following year with the appointment of a director, Alan Waterman from ONR, and barely enough money at the beginning to run an office. Within a few years, it will be supporting grants, fellowships, laboratories, equipment, and, for oceanography, even large research vessels. With ONR and NSF in operation, the two principal sources of oceanographic funding for the next half-century and beyond are now in place.

By the end of 1950, the WHOI fleet consists of *Atlantis* and *Caryn* and three smaller inshore boats, *Asterias*, *Mytilus*, and *Claire*. The Navy vessels have all been transferred, decommissioned, or taken out of service.

1951–1953

In 1951, *Atlantis*, *Caryn*, and *Albatross II* (a Fisheries vessel temporarily assigned to WHOI), respectively, spend 206, 199, and 161 days at sea. The 103-foot (31-meter) *Bear*, a former Army supply vessel, and the 50-foot (15-meter) dragger *Hazel III* are chartered for local work. *Bear* is purchased the following year. Early in 1952, *Atlantis* first loses its mizzenmast off the mouth of the Amazon and later the propeller shaft breaks, but following repairs, the vessel makes its first venture into the South

Three *Atlantis* captains—Adrian Lane, right, John Pike, center, and Scott Bray. Lane was master of *Atlantis* from 1946 to 1952, then port captain for a year. Pike followed Lane as both *Atlantis* master and port captain, and Bray followed Pike as *Atlantis* master.

The WHOI fleet took to the air in 1952 when this navy PBY6A was assigned to support the meteorological group. It is shown between missions at its base on Cape Cod's Massachusetts Military Reservation, not far from Woods Hole.

Atlantic for a study of western margin currents. WHOI contracts with ONR to supervise predesign studies of "superior oceanographic ships," an enterprise that will take several years. There is growing support for marine meteorology, and part of the WHOI fleet becomes airborne in 1952, when the Navy lends a PBY6A aircraft for research on trade-wind clouds and the role of sea salt nuclei in the formation of rain.

Funding during this period comes largely from the Navy (Bureau of Ships, Office of Naval Research, and Bureau of Aeronautics), with a substantial contribution to shellfishery investigations from the Commonwealth of Massachusetts. WHOI's first grant from the National Science Foundation supports newly minted Ph.D. John Ryther's comparison of algae thriving on duck farm waste in Long Island's Great South Bay with algae in similar but unpolluted waters. Ryther later describes the project:

> [T]he townships of Islip and Brookhaven in New York . . . requested that the institution investigate the condition of Great South Bay and the contiguous embayments on the south side of Long Island to determine the cause of the decline and collapse of what had once been a very prosperous oyster industry. The two towns each contributed the munificent sum of

$1,600 for the study. Grants of any size or kind in our field were so rare in those days that Dr. Redfield jumped at the opportunity. A group of six or maybe more WHOI personnel spent several months carefully studying the chemistry and biology of the bays, documenting for the record what everyone already knew. The tributaries to the bays were virtually lined with duck farms, and the . . . bay water looked like pea soup from the blooms of tiny green and blue-green algae that had displaced the normal phytoplankton flora. . . . [In] early fall of 1951, I read a small announcement in the newspaper that a new organization had been created in Washington—the National Science Foundation—and it was accepting proposals for grants. I quickly applied for and received [$3,000 for] what I believe to have been the first NSF grant to WHOI, to study the "etiology of the phytoplankton blooms of Great South Bay." [90]

During this period, Ryther, along with Ralph Vaccaro, is also examining methods for calculating ocean productivity (the rate at which energy is absorbed from sunlight for the synthesis of organic matter).

The Fay property barn is converted to a summer dorm offering three apartments and eight "bachelor quarters." Following long, unsuccessful negotiations with the Marine Biological Laboratory for the garage property, the Navy takes the land and begins construction on a 28,000-square-foot (2,520-square-meter) laboratory.

Athelstan Spilhaus chairs the 1952 Scientific Advisory Committee, which finds that more senior scientists are needed but that lack of staffing standards makes it difficult to attract them: "At the present time there is no policy on the length of contract or tenure . . . no policy on promotion . . . no policy on the termination of services in the case of inability to secure promotion." The committee also notes a need for education in oceanography. [91]

A new practice of inviting groups of local people to "inspect" the laboratory grows into an organization called Associates that is incorporated in 1952. Gerard Swope Jr. is elected president. By the following year he has a ten-member executive committee

Early meetings of the WHOI Associates Committee were held in the wardroom of *Atlantis,* including this one with Director Smith at left; Gerard Swope Jr., first president of the Associates, at right; and Jack Gifford, first secretary of the Associates, second from right.

Captain Bill II was chartered in the 1950s for Bill Schroeder's use in collecting fish. Here he displays two deep-sea fish called chimaera. Bigelow and Schroeder's revised, 577-page *Fishes of the Gulf of Maine* was published in 1953, followed the next year by the similarly weighty *Fishes of the Western North Atlantic: Sawfishes, Guitarfishes, Skates, and Rays.* Schroeder discovered a deep-sea lobster population that was being exploited by the mid-1950s.

that includes author Rachel Carson and future WHOI board chair Noel McLean. The fifty-member group makes plans to invite industrial corporations and institutions to become Associates. In 1956, gifts from the Associates program will total more than $78,000.

Visitors to the Institution in 1953 hail from Japan, France, Sweden, Canada, England, South Africa, Australia, India, Korea, Germany, Ireland, Thailand, and Iceland. They include French navy commandant Jacques-Yves Cousteau and His Imperial Highness Akihito, crown prince of Japan. A rare deep-sea fish caught by Bill Schroeder is presented to the prince as a gift for his father, who is well known for his interest in marine biology. Twenty-two years later, the emperor himself will visit Woods Hole laboratories.

General Foods becomes the first WHOI Corporate Associate in 1954 and donates its former marine base in East Boston to the Institution, along with $15,000, the largest gift since the original Rockefeller Foundation grant. By the end of the year, Corporate Associates include Edo Corporation, the International Nickel Company, Minneapolis-Honeywell Regulator Company, Raytheon Manufacturing Company, Shell Development Company, Standard Oil Development Company, Union Carbide, and Carbon Research Laboratories.

Mary Sears

Mary Sears probably knew the Woods Hole Oceanographic Institution better than anyone else during the first half of its existence. As a Radcliffe College graduate student and assistant to Henry Bigelow at Harvard's Museum of Comparative Zoology, her desk was near Bigelow's while he was writing the National Academy report that resulted in the founding of WHOI. During Woods Hole summers in the 1930s, she worked both with him and on her own plankton projects, and he generally saw that she got to know the world's leaders in marine science. She completed her Ph.D. in 1933, held the title of WHOI planktonologist beginning in 1940, and was named senior scientist in 1963 and scientist emeritus in 1978. In addition, she served tirelessly as clerk of the Corporation from 1947 to 1974.

Columbus Iselin observed that Mary Sears had "done as much for the advancement of oceanography as anyone I know." [92] As a WAVES (Women Accepted for Volunteer

Emergency Services) officer from 1943 to 1946, she ran the Oceanographic Unit of the Navy's Hydrographic Office, directing a staff of fifteen in the application of oceanography to war by assembling reports, charts, and pamphlets that served both Navy decision makers and operational units. She made her most enduring contribution to oceanography as editor of several journals and books, as well as many of the Institution's annual reports and *Collected Reprints* volumes. She almost single-handedly organized the First International Congress on Oceanography held at the United Nations in 1959. Twenty years later, she co-edited *Oceanography: The Past,* the proceedings of the Third International Congress on the History of Oceanography held in Woods Hole in 1980 as part of WHOI's fiftieth-anniversary celebration.

Because women were not permitted to go to sea until very late in her career, Sears's only research voyages took place aboard Peruvian fishing vessels during studies of plankton and El Niño conditions for a guano company in Lima (and it was she, not the men aboard, who dove into the sea to retrieve a plankton net that had parted from the line!). Recognizing the results of this work, Director Smith wrote in

In this 1960 photo, Mary Sears is surrounded by papers and biological samples in her Bigelow Laboratory office.

the 1953 annual report that Sears was "well recognized as a marine biologist, and . . . now qualified as a significant contributor to general oceanography. It may be noted with regret that present conditions do not permit her to carry out her work at sea in person except through engagements with foreign agencies." This was, of course, because there were no "proper facilities" for a lady. Sears later commented "But there were pails!"[93] and maintained that, though there were no proper facilities on the Peruvian vessels, what made the difference was that there were proper gentlemen aboard. Despite the frustration of not being allowed to go to sea, Mary Sears made the best of her situation.

In 1963, she christened *Atlantis II,* a WHOI ship with "proper facilities," and it was more than fitting that, in the year 2000, the U.S. Navy chose to name its sixth Pathfinder-class oceanographic survey ship USNS *Mary Sears.*

The Laboratory of Oceanography, shown in the foreground in an artist's rendering, was built by the Navy for WHOI's use in 1953–54. It was later named for the third director, Admiral Smith.

The principal developers of the salinometer—Karl Schleicher, right, and Alvin Bradshaw—are at work with their first model, in the mid-1950s, in the main lab of the research vessel *Bear*. Their salinometer was based on an instrument developed by the Coast Guard for ice patrol work. It replaced silver nitrate titration, eliminated problems of using glassware at sea, improved the accuracy of salinity measurements, and doubled the number of analyses that could be handled in an hour to about twenty.

1954–1956

The nuclear age comes to WHOI with the arrival in 1954 of geochemist Vaughan Bowen from Brookhaven National Laboratory. He studies the distribution of long-lived fission products in present-day Atlantic seawater and uses radiochemical techniques for detecting substances present in such small amounts that other analytical methods don't register them. WHOI hosts a conference in August 1954 for representatives of the principal agencies concerned with disposal of radioactive wastes to consider what research would help determine whether disposal at sea is feasible.

In June 1954, the Institution dedicates its second laboratory building, the Laboratory of Oceanography (later named Smith Laboratory), constructed by the U.S. Navy both to house classified work and to relieve overcrowding. Following the ceremony, laboratories as well as the research vessels *Atlantis* and *Bear* are opened for inspection (along with Stan Eldredge's carefully labeled "Scientific Carpenter Shop"). A three-day convocation on physical, chemical, and

biological oceanography; marine meteorology; geology; and geophysics begins. *Oceanus* magazine boasts that "the convocation was attended by almost all the foremost oceanographers of the western world."[95]

Progress in instrumentation includes retrievable moorings—Dave Frantz manages to reduce the explosive charge needed to activate an anchor-releasing relay from one hundred to two pounds! Karl Schleicher and Al Bradshaw build the first WHOI salinometer, taking advantage of the fact that the ability of seawater to conduct electricity depends on its salinity and temperature. Easier and quicker than chemical titration, use of the salinometer increases the precision of salinity measurements by about five times and means that salinities for a cruise can be complete when the ship returns to port.

Atlantis makes its first trip to the Pacific in 1955 to collect sediment cores off the west coast of South America. Chief scientist Henry Stetson dies at sea during this cruise, and Parker Trask directs the remainder of the cruise. While collecting one hundred sediment cores between 4° and 25° south, the scientific party also makes the second broad-scale systematic survey of the Peru-Chile trench, using Bud Knott's new echosounder, called the precision graphic recorder. Trask speculates that the trench's deeps were first identified by cable-laying ships as early

Atlantis transits the Panama Canal in 1955, en route to its first excursion to the Pacific Ocean to collect sediments and conduct bottom surveys off the west coast of South America.

Biologists Alfred Redfield, top, and Bostwick Ketchum, who were professor and student, respectively, in the 1930s, worked together at WHOI for many years. They served in similar administrative positions both at WHOI and outside the Institution.

as 1875 and notes that the first expedition to undertake "anything like a systematic exploration of the trench" was a 1952 Scripps cruise. *Atlantis* data reveal contours as deep as 25,000 feet (7,576 meters).[96]

In 1955, the American Society of Limnology and Oceanography (ASLO) elects Alfred Redfield as president and reelects Buck Ketchum as secretary of the organization. These two scientists maintain a long-time close association, which began in the 1930s when Ketchum was a graduate student with Redfield at Harvard and they made summer research cruises aboard *Atlantis*. Both take the broad view of oceanography advocated by Bigelow and are equally at home in the physical, chemical, and biological realms. (A quotation from Redfield—"Life in the sea cannot be understood without understanding the sea itself"—is inscribed on the dedication plaque for the WHOI laboratory that bears his name.) Both were ecologists before the term was popular. Redfield and Ketchum not only work together as scientists but also take similar organizational paths. Ketchum becomes ASLO president in 1959; each has a term as president of the Ecological Society of America, Redfield in 1946 and Ketchum in 1955; and each serves as WHOI associate director, Redfield from 1942 to 1956 and Ketchum from 1962 to 1977.

Stan Poole records the first underwater sounds of baleen whales from *Risk* when three right whales spend a week in Menemsha bight off Martha's Vineyard. These sounds join the growing records from toothed whales collected by Bill Schevill.

When Smith steps down as director in August 1956, Iselin "reluctantly" becomes director again because he believes the International Geophysical Year, planned for 1957–58, is important to the future of oceanography.

A proponent of clean beaches and better recreational facilities, Dr. Redfield, together with Dr. Bostwick H. Ketchum, attacked [the study of pollution in harbors and estuaries] with such vigor and novel approach that the results have caused great interest among engineers having to deal with the flushing of harbors and among scientists studying the mixing of superimposed layers of fresh and salt water in such structures.[97]

—*Oceanus* magazine, 1953

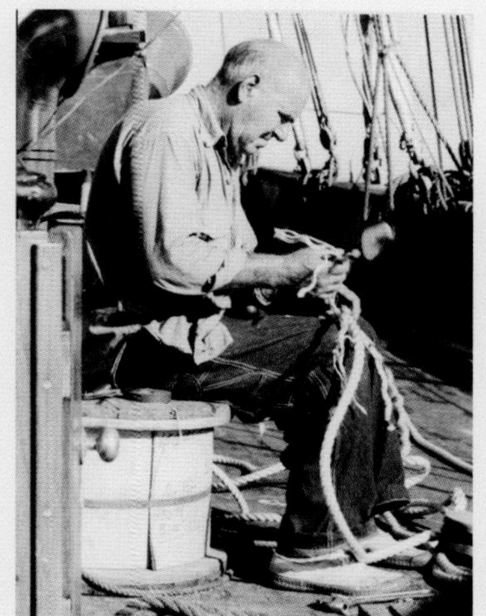

Captain Arvid Karlson has sailed away. . . . [T]he Institution has lost a valuable member. . . . [H]is ashes will be carried out to sea, which was his home for 50 years. . . . There are but few, regardless of rank, of whom it may be said: "He was a real seaman." Captain Karlson belonged to that vast disappearing race of men whom we generally think [of] as clipper ship or whaling captains. Born in Vetlanda, Sweden, he left home at the age of 14 to sail on square riggers, until the age of steam caught up with him. . . . A man of few words, but those well chosen, his remarks became legendary. . . . Captain Karlson . . . sailed for many years as first mate on the R.V. Atlantis *under Captain Adrian K. Lane, until receiving command of the R.V.* Caryn. *His years of care of the* Atlantis *show unto this day, and it is largely due to Karlson's efforts "to keep up ship" that the vessel is still able to work and work hard. Unexcelled in celestial navigation, the stars were his closest friends. Let those of us who look at them, remember him. Fair winds, Skipper.*[98]

—Jan Hahn, *Oceanus* magazine, 1955

Arvid Karlson spent half a century at sea, including ten years aboard WHOI ships. He joined *Atlantis* in 1945, was first mate for several years, and then took command of the sailing vessel *Caryn*.

The science party ready for action on *Atlantis* Cruise 215 (Woods Hole to Bermuda to Nassau to Woods Hole) included, from left, Gerry Metcalf, Conrad Malicoat, G. B. Ferguson, Alan Faller, Val Worthington, and Peter Hughes. Metcalf and Worthington not only spent a great deal of time at sea in temperate climes for studies in physical oceanography, but both were also involved in Arctic and Antarctic expeditions. Metcalf joined the WHOI staff in 1946 following discharge from the Navy. He spent part of each of the next ten years assisting in Navy polar hydrographic studies, first during Operation Highjump in the Antarctic and then in the Arctic during surveys taken for early-warning systems that would sound the alarm in the event of a Soviet attack over the pole.

Crawford, a 125-foot (38-meter) Coast Guard cutter acquired in 1956, "after plowing through a mountain of red tape,"[99] brought greater range and speed to the WHOI fleet. Here the ship leaves on its first Institution cruise, bound for studies in the tropical birthing area of hurricanes. After disruptions from several major storms in the mid-1950s, and building on ten years of WHOI research into the dynamics and physical chemistry of warm cumulus clouds characteristic of the marine tropics, the scientists were hoping to find a "weak link" in hurricane formation. (They didn't.)

Columbus O'Donnell Iselin, Twice Director

While Henry Bigelow and Frank Lillie loom large in the Institution's prehistory and first years, Columbus Iselin is most responsible for its operational strength at sea from the very beginning, its early excellence in physical oceanography, its wartime management, and its survival through the rocky postwar years. A strong friendship with Bigelow no doubt lent him courage and conviction, but Iselin himself was the most important continuing influence on the Woods Hole Oceanographic Institution from 1930 through the early years of Paul Fye's tenure as director. When Iselin died in 1971, Fye called him "one of the giants of oceanography" and said, "he was my constant mentor and advisor" in the last decade of his life.[100]

Columbus Iselin sees Edmund Watson and others off on a 1950 Gulf Stream cruise aboard a boat called *Seal*. Iselin appears happiest in photographs taken on or near the sea, and he liked to conduct business on the WHOI pier whenever possible.

As director during the war years, Iselin looked after the Navy's needs but also nurtured work that would continue when WHOI scientists' attention could turn again to more academic questions. He traveled the halls of Washington, keeping WHOI at the forefront of naval needs during the war, while at home maintaining the threads of basic research. His management style was casual but involved. "Even when the Institution grew to include several hundred people, Columbus remained personally and technically involved in individual problems to an amazing extent," Allyn Vine wrote. "He spent a great deal of time wandering around the lab, quietly observing, gently encouraging, and posing questions so pertinent that they frequently suggested solutions." [101]

With his long military background, Edward Smith never quite adjusted to the Institution's can-do but unstructured atmosphere. Henry Stommel wrote, "In many ways, during Smith's term in office (1950–56), Columbus remained a surrogate director. People automatically referred their ideas and plans to him. . . . And after Smith's departure, Columbus returned for two more years as director. One could hardly notice the difference except in salaries and that the halls had fresh coats of paint." [102]

Iselin wasn't perfect—fiscal management was not his strong suit, and, as one employee put it, he didn't need money, so he didn't quite understand that others did. Living on Martha's Vineyard and in Milton, Massachusetts, Iselin generally left Institution-related entertaining to those who lived in Woods Hole. But he was surely the right man for this period in WHOI's history. Raymond Stevens, chairman of the board from 1955 to 1960, said, "Columbus commanded respect wherever he appeared, in whatever company. With larger ships, with expanding programs, and particularly increased specialization it will be a rare man, if any, that ever matches his universal coverage of the science and the art of ocean research, and of seamanship or his familiarity with the total literature, the researchers themselves—and, not least, the sea itself.

"Whenever I saw Columbus near a boat, I had the feeling that he was a true seaman—the man, the boat or ship, and water seemed a unit, belonging together." [103]

In 1959, *Time* magazine chose Columbus Iselin to represent U.S. oceanography on its cover.

Iselin chats with Navy officials during a visit to address staff and students at the Naval War College in Newport, RI, in 1961.

1957–1958 In 1957, about one-sixth of the U.S. budget in oceanography ($1.87 million) comes either directly to WHOI or to support cooperative programs with the Lamont Geological Observatory of Columbia University, Texas Agricultural and Mechanical College, Scripps, the University of Washington, and several government facilities. Nearly half of the Institution's effort concerns large- and small-scale ocean circulation, using the basic measurements of temperature, salinity, and dissolved oxygen as well as mathematical modeling. The modeling group meets biweekly with a similar group at MIT. Another program analyzes drift bottle data as background for setting out radio-telemetering buoys. Buck Ketchum leads an expanding program in marine biology funded by the Atomic Energy Commission. It includes the beginning of a three-year study of continental shelf plankton populations and new work by John Ryther and Charles Yentsch on calculating productivity.

The International Geophysical Year (IGY) begins in 1957; at the same time, Iselin reports a funding crisis for the Woods Hole Oceanographic Institution. He writes in the annual report, "During the past year we have had some notable scientific successes . . . but we have also been going through some very rough weather financially. As the year progressed funds available through the military services for the support of basic research steadily declined." Though IGY funding kept ships and scientists at sea, Iselin noted that "I do not remember a year when the budget involved so much worry and uncertainty. The future would look most uncertain indeed were it not for the establishment of a new Committee on Oceanography by the National Academy of Sciences–National Research Council." Iselin was one of ten people appointed to the committee, and its list of "Members, Panel Chairmen, and Staff, November 1957 through July 1962" also includes Paul Fye and Allyn Vine.[105]

Bill Schevill, left, founded the field of marine mammal bioacoustics just after World War II. When Bill Watkins, right, joined him at WHOI in 1958, they began what current WHOI bioacoustician Peter Tyack calls "a 40-year collaboration that changed the face of marine mammal science." Watkins responded to the problem of how to record the animals' sounds from small vessels by building a portable tape recorder; he went on "to develop most of the scientific tools that form the basis of our field," Tyack says.[106]

The USSR launches the first two *Sputnik* satellites in October and November 1957, while the first U.S. satellite is still being readied for launch. Though it will take a while to translate into dollars, science gains new national priority.

By the end of the war our annual operating expenses had increased from roughly $150,000 per year, the approximate income from endowment, to about $1,000,000 per year. The Federal government had suddenly become a major source of support for our oceanographic laboratories. . . . There is worry that this form of support may not remain sufficiently stable over the years to build up a strong scientific staff, but to date, with few exceptions, the policy in Washington has been remarkably liberal and understanding. Meanwhile, our own uncommitted income has grown to about $300,000 per year. At the present time this is about 10 per cent of the total cost of our operations, but it is the key to their success.[107]

—Columbus Iselin, unpublished "Short History of the Woods Hole Oceanographic Institution," about 1960

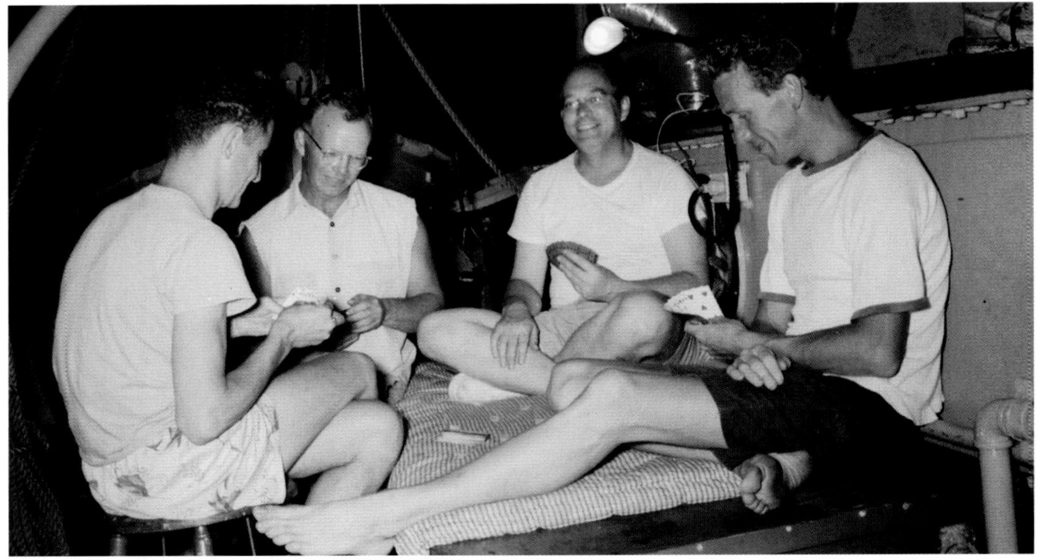

WHOI welcomes a new director in June 1958 when Paul Fye takes the helm, and Iselin becomes Henry Bryant Bigelow Oceanographer. Fye writes in his annual report that although the budget for 1959 is only 10 percent higher than 1958's $3 million, there is increasing interest from the defense industry in physical oceanography and a common opinion in Washington that oceanography will expand significantly in the near future. "The only question appears to be how much and by what means. One of the most important decisions the Institution has ever had to make is what role it should play in this expansion." This leads to the question of how much should the plant and staff of the Institution grow, something the scientific staff had been thinking hard about for some time.[108]

Henry Stommel and the Global Ocean Circulation

Henry Stommel is generally considered to be the leading physical oceanographer of the twentieth century. A conscientious objector, he was in his second year as an astronomy graduate student at Yale when he came to WHOI in 1944 as a way of serving his country without going to the battlefield. He would maintain close, collaborative ties with WHOI colleagues throughout his career though he held faculty appointments at Harvard from 1960 to 1963 and MIT from 1963 to 1978, when he returned to the Institution. Colleagues tend to describe a scientific conversation with Stommel as a situation where they had to "hang on" to follow his intellectual gymnastics, but they also laud his generosity with ideas (he had more new ones that he could follow up on himself) and his spirit of collegiality as an inspiration to "legions of oceanographers."[109]

Their tributes to him are collected in a special commemo-
rative issue of *Oceanus*
magazine (vol. 35, 1992).

Stommel took the first
important step toward an
understanding of the general
circulation in 1947 in
response to a question posed
by Iselin. Why, Iselin wanted
to know, does the Gulf Stream

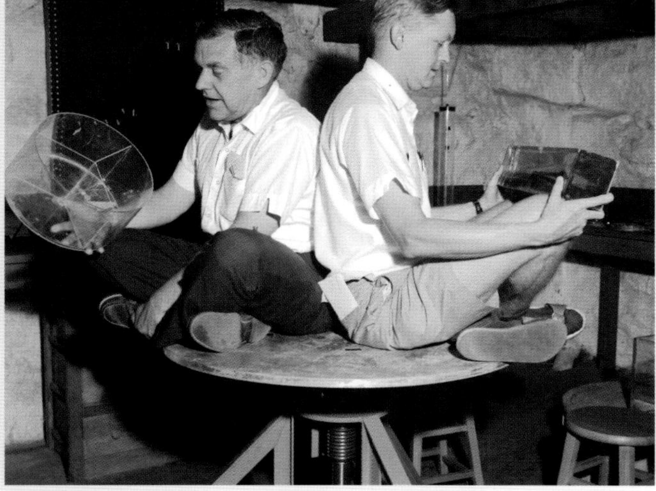

As principal instructors for the 1968
Geophysical Fluid Dynamics program on
the theme "general circulation of the
ocean," Henry Stommel, left, and Lou
Howard (both on the MIT faculty at the
time) hopped onto a turntable in the
basement of Walsh Cottage to demonstrate the effect of rotation on seawater.

flow as a narrow current on the western side of the Atlantic? Working with a simple
model, an approach that would become his trademark, Stommel identified the first of
the two major driving forces in ocean circulation. He determined that the rotation
and curvature of the earth are the essential factors producing intense currents on the
western sides of all the oceans of the world. This was the first realistic dynamical
model of ocean circulation.

Then, in 1955, Stommel theorized that there must be a deep, thermohaline-driven
countercurrent under the Gulf Stream. He identified the second major driving force
for ocean circulation when he showed that changes in density caused by temperature
(thermo) and salinity (haline) can be responsible for deep flows in the ocean.

At the same time, John Swallow of the British National Institute of Oceanography
was developing neutrally buoyant floats that could follow deep currents. A March 1957
British-American *Discovery II–Atlantis* expedition both definitively tested Swallow's
floats and found that Stommel's deep western
boundary current indeed existed.

Almost all previous theories of general ocean
circulation considered currents that could be modified by density differences but that would not be
significantly influenced by salinity and temperature
changes. It was now clear that thermodynamic
forces must be as important as the wind in setting
up the large-scale current systems in the ocean.
Once the countercurrent was established beyond a
doubt, other facts and theories about global circulation began to fall into place.[110]

Rigorous intellectual discussion and free
exchange of ideas were Henry Stommel's
hallmarks. He is shown, in 1979, in a discussion with long-time colleague George
Veronis, left, of Yale University.

The Bodman (or Bowen) bottle was the second generation of large-volume water samplers. The heavy bottle was a beast to handle, as this group is doing aboard *Crawford,* probably in the late 1950s.

A thorough 1958 inspection of *Atlantis* indicates that it would cost as much as the original construction to operate the ship beyond 1960. Consequently, a ship design committee forms under the direction of Francis Minot, one of the original designers of *Atlantis.* In addition, Allyn Vine is dispatched to work with the Navy on conversion of existing ships for modern oceanography. He and Brackett Hersey help persuade the Navy to release two salvage and rescue ships from the mothball fleet, assigning *Chain* to WHOI and *Argo* to Scripps. WHOI has already gained some larger-ship experience by using the 179-foot (54-meter) Coast Guard cable layer *Yamacraw* for eleven cruises beginning in 1957. The geophysics group transfers its cumbersome 600-foot (182-meter) thermistor array from *Yamacraw* to the 215-foot (65-meter) *Chain* in Savannah in the fall of 1958 for work on the Blake plateau during *Chain's* first WHOI cruise. With *Chain's* arrival, the large research vessel era dawns for the Woods Hole Oceanographic Institution.

It is also a new age for WHOI ashore. "In accordance with the Trustees' directive . . . that the Institution plan for further orderly expansion while maintaining the high excellence of its scientific program," the director establishes committees for land acquisition, a building program, and educational policy in addition to the research vessel design committee.[111]

The International Geophysical Year (IGY)

Oceanography was one of eleven earth sciences emphasized during the International Geophysical Year (the others ranged from "aurora and airglow" to "seismology and solar activity").[112] The International Council of Scientific Unions proposed IGY in 1952, with forty-six countries originally agreeing to participate in a wide range of experiments and observations; by the close of the activity, sixty-seven countries were involved. Oceanographic areas of emphasis were (1) short- and long-term sea level change; (2) deep-water motions, especially in equatorial and subpolar regions; and (3) multiple ship operations across the boundary of temperate and polar waters and at the outflow of Mediterranean water into the North Atlantic. *Atlantis, Crawford,* and *Chain,* often working in conjunction with Great Britain's *Discovery II,* were among the sixty ships involved in IGY research cruises from fall 1956 through mid-1959 (though the official IGY dates were July 1, 1957, to December 31, 1958).

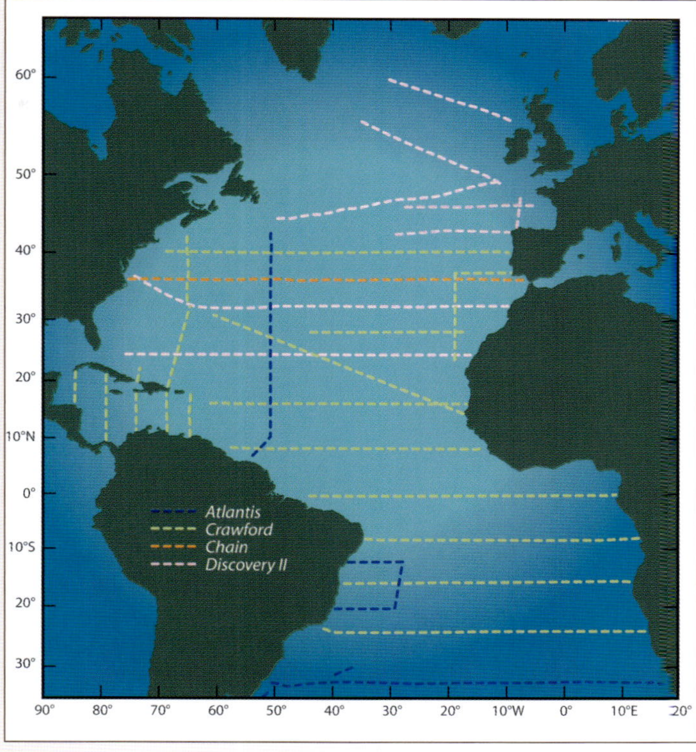

The map above shows Atlantic IGY sections made by *Atlantis, Crawford, Chain,* and *Discovery II* (UK), minus transits to and from ports. On the 24°S section, after the winch broke down two days out of Rio de Janeiro, *Crawford* scientists and crew worked fifteen straight hours in the hot equatorial sun and through the humid night to retrieve some ten thousand feet of wire carrying eleven Nansen bottles.

Dick Leahy, a research associate in geology and assistant to the director, coordinated the Institution's IGY activities, which at one time or another involved a majority of those working at WHOI. Columbus Iselin was in the thick of worldwide IGY planning, noting at one point that "we are in danger of having more international committees in oceanography than there are qualified people with the time to serve on them." Concluding many years of patient effort, WHOI scientists perfected deep-sea hydrography in order to collect data from the top to the very bottom of the water column, and they confirmed the Atlantic's deep western boundary current (see page 97). Two atlases, one for the upper 500 feet (152 meters) of water and the other for the deep ocean, compiled IGY hydrographic data and echosounder profiles for the Atlantic Ocean. Vaughan Bowen's large-volume Bodman water sampler (named for Ralph Bodman of the technical services staff) had its first use on *Crawford*'s equatorial section. In addition to the Atlantic sections plotted on the map, the Institution was involved in IGY hydrographic work in the Mediterranean and Red seas as well as the Gulf of Aden, research at two Arctic Ocean ice stations, flights for atmospheric carbon dioxide sampling, and cooperation with a multitude of foreign investigators.

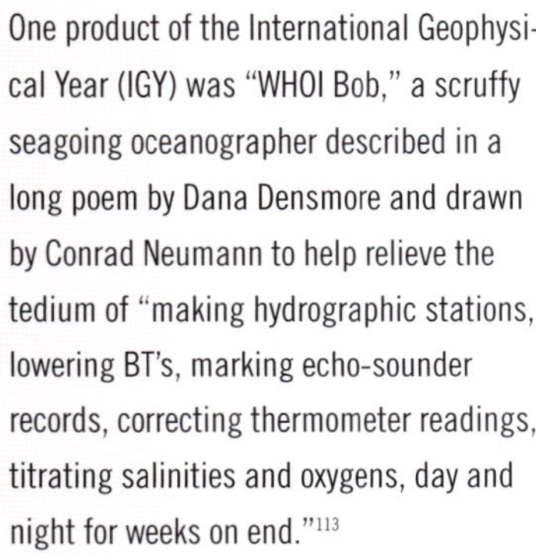

One product of the International Geophysical Year (IGY) was "WHOI Bob," a scruffy seagoing oceanographer described in a long poem by Dana Densmore and drawn by Conrad Neumann to help relieve the tedium of "making hydrographic stations, lowering BT's, marking echo-sounder records, correcting thermometer readings, titrating salinities and oxygens, day and night for weeks on end."[113]

Pioneering Chemical Oceanography:
A Short History of WHOI Research in Marine Chemistry

Norris Rakestraw, an assistant professor of chemistry at Brown University, was among the earliest WHOI scientific-staff members. He and his colleagues kept the Institution in the mainstream of pioneering chemical oceanography during its first decade. Their work included examining the seasonal variations of various substances in seawater at stations in the Gulf of Maine and on a line from Woods Hole to Bermuda. Studies of marine biological chemistry initiated by Alfred Redfield continued for many years and eventually yielded the Redfield/Ketchum/Richards ratios of carbon, nitrogen, and phosphorus in phytoplankton, particulate matter, and dissolved forms. These ratios are still applied today in studies of oceanic biogeochemistry.

During World War II and in the immediate postwar years, the chemical component of WHOI research focused largely on antifouling paints for ships and the chemistry of

Brown University professor and chemist Norris Rakestraw, shown here following a 1935 *Atlantis* cruise to the Gulf of Maine, was one of Henry Bigelow's earliest WHOI staff appointments.

explosives (see pages 61 to 65). With staff relocations, there was a fallow period for marine chemistry in the late 1940s, but it did not last long. Soon there was new work on oxygen minimum zones and their relationships to underlying sediments, as well as research interest that would last for decades on the inputs, fates, and effects of the oceanic disposal of various types of wastes.

Studies of fission products in seawater in the mid-1950s by Vaughan Bowen (see page 88) indicated that some isotopes were removed rapidly from upper layers of the sea and initi-

Vaughan Bowen, kneeling, worked with WHOI shop personnel beginning in the 1950s to develop large-volume water samplers for his work on fission products in seawater. Here, a first-generation sampler goes down under vacuum.

ated longstanding work on the transport of a variety of substances through the water column aboard descending particles. Bowen's laboratory became a world leader in investigations of fission-product radionuclides in the oceans.

Max Blumer joined the staff in 1960 to begin an ambitious program of extracting and isolating trace amounts of organic compounds from seawater and sediments that led to the Institution's prominence in marine organic geochemistry.

Henry Bigelow's view of oceanography as an integrated science was evident in the initial organization of science departments: chemistry was initially paired with biology in 1962 and then with geology the following year when petroleum chemist John Hunt became the first permanent department chair for these disciplines. A reorganization later in the 1960s created both a chemistry and a geology and geophysics department.

Focusing on major and minor elements in seawater, relationships between dissolved and particulate matter, and sediment and underwater-basalt geochemistry, the department was on its way to becoming one of the foremost chemical oceanography and marine geochemistry research efforts in the world. A technical staff that provides outstanding analytical expertise has been key to maintaining that leadership.

With growing interest in pollution, WHOI staff made some of the earliest open-ocean measurements of DDT and initiated an abiding interest in the persistence of oil spilled or otherwise transmitted to the marine environment (see page 130).

Research in today's Department of Marine Chemistry and Geochemistry encompasses all of the areas mentioned above and also includes, among other subjects, studies of exchanges at the air-sea interface, the ocean carbon cycle, microbial and planktonic processes, hydrothermal systems, paleoceanography, and shoreline groundwater.

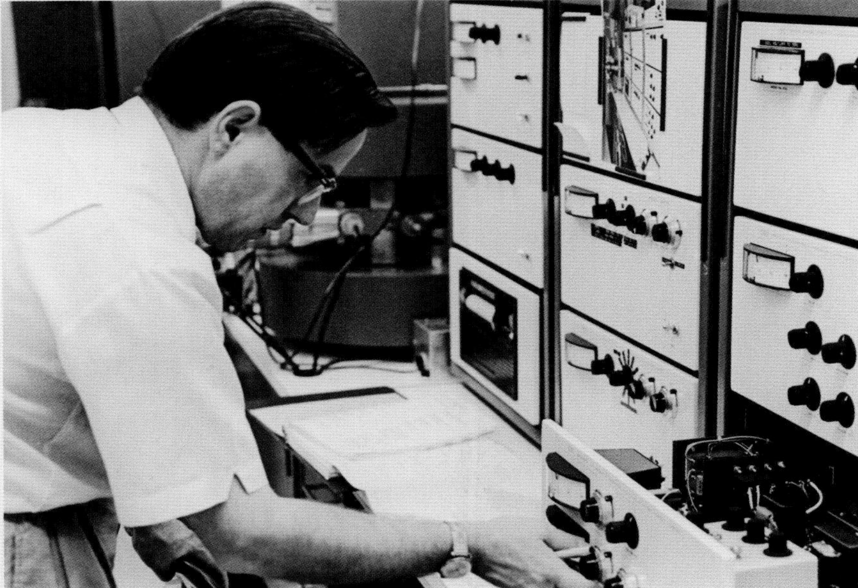

Max Blumer, who joined the WHOI staff in 1960, was a pioneer in detecting trace amounts of organic compounds in seawater and sediment.

This feature is based on an unpublished 2004 manuscript entitled "Chemical Oceanography/ Marine Chemistry and Geochemistry at WHOI" written by John W. Farrington, WHOI senior scientist and vice president for academic programs and dean.

Considering Marine Bacteria:
A Short History of WHOI Microbiology

Henry Bigelow asked Selman Waksman in the spring of 1931 whether he thought the new oceanographic institution should "consider marine bacteria." Waksman, a Rutgers University researcher, answered affirmatively and, after spending a month in Woods Hole the following summer, presented Bigelow with a plan for a program in marine bacteriology. Every summer from then until 1942, Waksman and several colleagues worked at WHOI. Waksman suffered terrible seasickness during a trip with Bigelow in 1932, so he was excused thereafter from Bigelow's maxim that all investigators should

Henry Bigelow invited Selman Waksman to develop the Institution's first program in marine microbiology in 1931. Here, in 1932, Waksman gets a bit of fresh air to relieve his seasickness during his only *Atlantis* cruise—after that he sent assistants to collect samples.

go to sea each year, and assistants went out on *Atlantis* in his place.[114] Many of the problems Waksman's group addressed are still topical today, including the nature of bacterial populations in seawater and sediments, bacteria involved in the nitrogen cycle, transformations of dissolved organic matter, and the agents responsible for wasting disease in eelgrass.

Though Waksman later served on the WHOI Corporation, by 1942 his attention was largely focused on the search for antibiotics. He was awarded the Nobel Prize in Physiology/Medicine in 1952 for the discovery of streptomycin, the first antibiotic effective against tuberculosis.

There was little activity in microbiology for the rest of the 1940s. However, in 1951, Alfred Redfield offered Stanley Watson, then at the University of Washington's Friday Harbor Laboratory, a one-year fellowship to renew the study of eelgrass wasting disease and its causative agent, *Labyrinthula.* Watson left in 1952 to pursue graduate studies at the universities of Washington and Wisconsin, but returned in 1957 to initiate investigations of marine nitrifying bacteria. Extensive studies over the next thirty years identified many novel nitrifiers and brought his laboratory recognition as a world authority on these unusual bacteria, which play a central role in marine and terrestrial

nitrogen cycles. Watson also had a second career in biotechnology after determining that the clotting system in the blood of horseshoe crabs that he was using to measure bacterial biomass in seawater might have a medical application. His success in marketing a product to screen for bacterial toxins in pharmaceutical products allowed him to become a significant WHOI benefactor.

Stanley Watson monitors the oxygen uptake of microbes in a seawater sample during a 1961 *Chain* cruise.

Watson met German microbiologist Holger Jannasch in 1959 and invited him to visit Woods Hole. Later, after completing advanced studies in Germany, Jannasch joined the WHOI scientific staff in 1963. Over the next thirty-five years, he and his WHOI colleagues made major contributions in three areas. First they studied the growth and ecology of bacteria at very low nutrient concentrations. Next they investigated the ecology of bacteria living under high pressure and at low temperatures (inspired by the remarkable preservation of lunches in the sunken *Alvin*—see page 121). Their third area of study involved one of the twentieth century's great discoveries in microbial ecology, that the abundant life surrounding the deep-sea hydrothermal vents is supported by bacterial chemosynthesis associated with sulfur oxidation.

Holger Jannasch works with a chemostat in his laboratory about 1966. The continuous culture apparatus was designed to allow Jannasch and colleagues to provide a continuous flow of nutrients to the marine sulfate-reducing bacteria they were studying.

This long history of bacterial research at WHOI continues today with investigations of various microbial processes in the ocean, including hydrothermal vent ecology, the role of cyanobacteria (photosynthesizing microbes) and other bacteria in the nutrient cycles of the ocean, and microbial dissolution and growth on deep-sea minerals.

This feature is based on an unpublished 2004 manuscript of the same name by WHOI senior scientist John B. Waterbury.

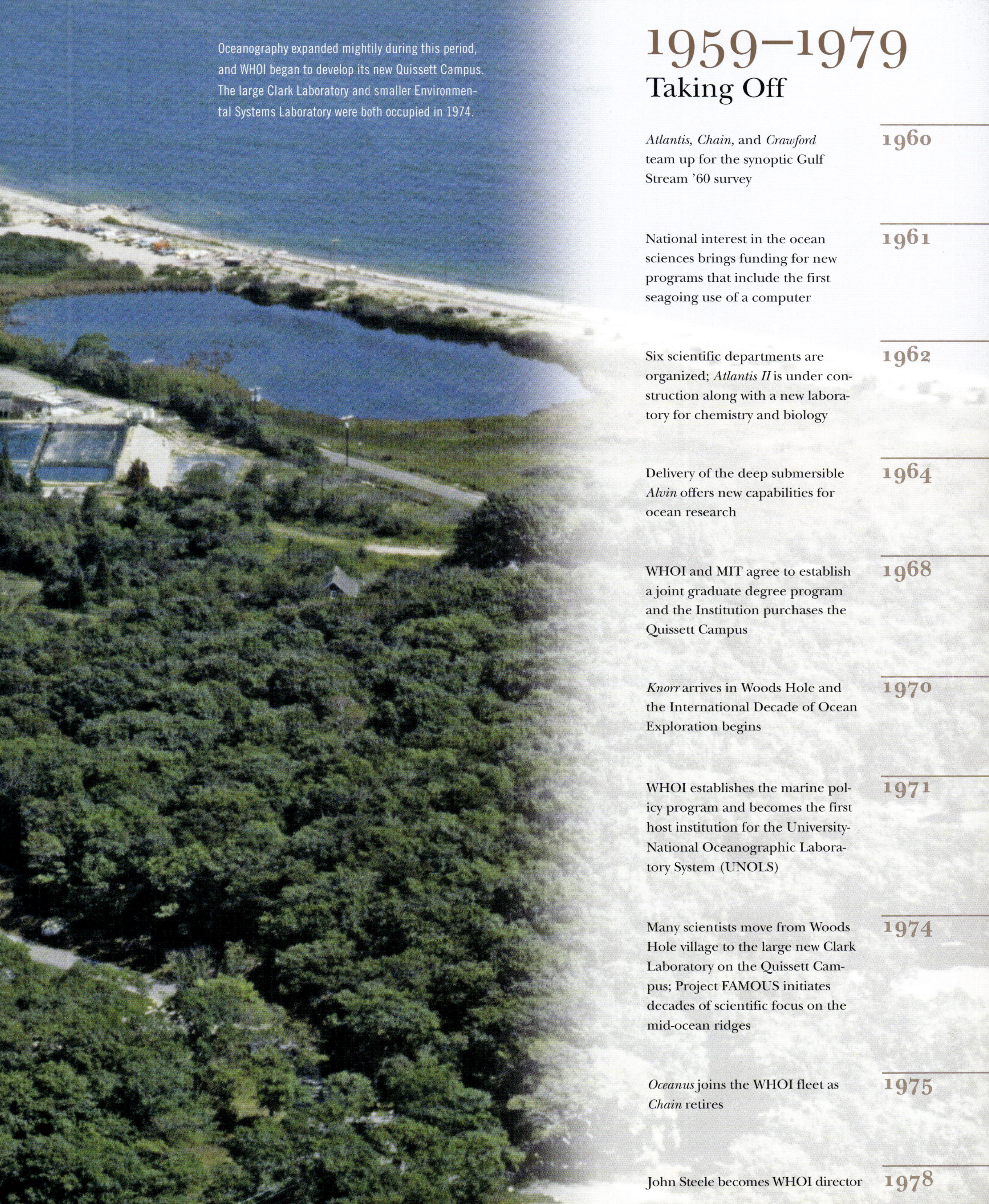

1959–1979
Taking Off

Atlantis, Chain, and *Crawford* team up for the synoptic Gulf Stream '60 survey — **1960**

National interest in the ocean sciences brings funding for new programs that include the first seagoing use of a computer — **1961**

Six scientific departments are organized; *Atlantis II* is under construction along with a new laboratory for chemistry and biology — **1962**

Delivery of the deep submersible *Alvin* offers new capabilities for ocean research — **1964**

WHOI and MIT agree to establish a joint graduate degree program and the Institution purchases the Quissett Campus — **1968**

Knorr arrives in Woods Hole and the International Decade of Ocean Exploration begins — **1970**

WHOI establishes the marine policy program and becomes the first host institution for the University-National Oceanographic Laboratory System (UNOLS) — **1971**

Many scientists move from Woods Hole village to the large new Clark Laboratory on the Quissett Campus; Project FAMOUS initiates decades of scientific focus on the mid-ocean ridges — **1974**

Oceanus joins the WHOI fleet as *Chain* retires — **1975**

John Steele becomes WHOI director — **1978**

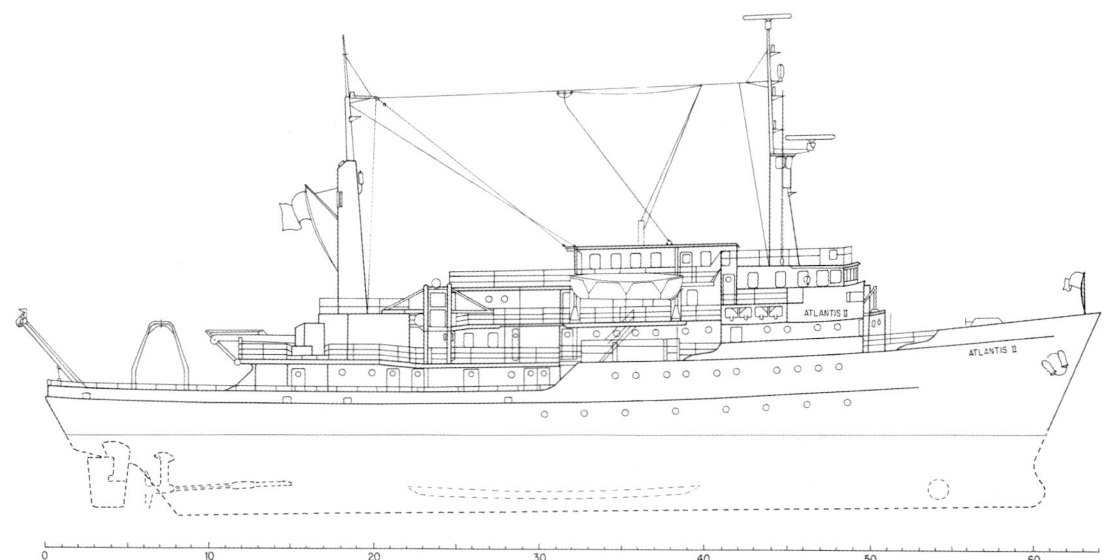

Years of planning went into the design for *Atlantis II*, whose name was quickly shortened to *A-II*, while *Atlantis* was known as "the A-boat." The new ship was intended to "do effective all weather oceanographic research from the fringe ice to the tropics and…accommodate more than one scientific discipline on a given cruise." The planned 170-foot (52-meter) vessel grew to 210 feet (64 meters) with the decision to make it steam-powered for quiet operation. (A steam engine requires more fuel than a diesel engine, hence, a larger ship was required.) This drawing is dated 1962, when the ship was in the final stages of construction for delivery that December.

1959–1961

Prospects for oceanography continue to improve. In January 1959, Admiral Arleigh Burke approves the Navy's "Project TENOC: The Next Ten Years in Oceanography," and in February, the National Academy of Sciences/National Research Council releases the report "Oceanography 1960–1970." TENOC makes a case for Navy support of basic research and recommends tripling the budgets of the civilian oceanographic institutions. The NAS report calls for ambitious educational, research, sea-going, and institutional programs. NSF awards WHOI $3 million for construction of the first research ship to be built in the 1960–70 program. In March 1961, President Kennedy requests an increase in funding for oceanographic research from about $60 million to nearly $100 million for fiscal year 1962. He tells the U.S. Senate, "Knowledge and understanding of the oceans promise to assume greater and greater importance in the future. This is not a one-year program or even a 10-year program. It is the first step in a continuing effort to acquire and apply the information about a part of the world that will ultimately determine conditions of life in the rest of the world." [116]

> *"Together let us explore the stars, conquer the deserts, eradicate disease, tap the ocean depths and encourage the arts and commerce."*
>
> *"We have neglected oceanography, saline water conversion, and the basic research that lies at the root of progress."* [115]
>
> —President John F. Kennedy, Inaugural Address and State of the Union Message, 1961

The eight-week, NSF-funded summer program "Theoretical Studies in Geophysical Fluid Dynamics" holds its first session in 1959. Topics considered by eleven staff and four graduate students center on the fundamental theory of rotating and stratified fluids. The course is considered a "most successful venture," and plans are drawn for a second session in 1960. Over the years, the program will come to be known simply as "GFD," and 2005 will mark its forty-seventh anniversary.

In late summer 1959, Paul Fye transfers WHOI headquarters to New York City for the First International Congress on Oceanograpy, organized by Mary Sears at United Nations headquarters. He calls the August 21 to September 12 meeting the "highlight" of 1959. Seventy WHOI staff are among the twelve hundred oceanographers attending the meeting from around the world (thanks to a daily shuttle from Woods Hole to New York City run by the Institution's current aircraft, an R4D). *Chain* and *Atlantis* call at New York's Pier 34 and welcome nearly a thousand visitors from twenty-seven countries during the congress.

R/V *Bear* participates in 1959 with three other oceanographic ships in one of the first multiship seismic surveys undertaken by U.S. scientists. They focus on an area north of Puerto Rico, looking for a possible site to drill a borehole to the Mohorovicic discontinuity, the boundary between Earth's crust and mantle, which is closer to the surface beneath the oceans than it is under the continents. "Project Moho" doesn't itself succeed, but it provides a foundation for deep-sea drilling programs that will continue into the twenty-first century.

For many of its forty-seven years, the summer geophysical fluid dynamics (GFD) program has met in Walsh Cottage, the porch of which has been the scene of myriad sophisticated GFD discussions. Each year, the participants gather on the porch for a group photo, including this one from 1968.

Beginning in 1952, aircraft were used regularly for nearly twenty years by WHOI researchers for meteorological, marine mammal, changing coastline, and other observations. A 1958 issue of *Oceanus* magazine noted that Claude Ronne, who held the title "research associate in photography," had logged more than a quarter of a million air miles.[117] The first plane, the PBY6A pictured on page 84, was followed by a Stinson Voyager, a DC-3, a DC-4, and a C-54Q, along with this Helio Courier acquired in 1959. Some were loaned by the U.S. government and some were purchased. The hangar (behind the plane in the photo) and launching ramp on Dyer's Dock off Water Street allowed convenient use of the amphibious Helio Courier. WHOI's airborne fleet ended operation in 1970, when the Helio Courier was sold and the C-54Q was returned to the Navy.

Women Pioneers at Sea

The story of women at sea on WHOI ships began quietly on April 8, 1952, when the husband-and-wife team of Harvard biologists Barbara Lawrence and William Schevill, who was also a WHOI research associate, joined R/V *Caryn* in St. Georges, Bermuda, for a week of research on whale sounds. Lawrence's work at sea was not widely known, and in the early 1950s it was generally thought that women simply did not go to sea—indeed, some believed they would bring a ship bad luck.

The story continues with a stowaway. Radcliffe graduate student Roberta Eike had summer fellowships in 1955 and 1956 to study marine crustaceans with George Clarke in Woods Hole. On September 15, 1955, she wrote a short document that generally objected to the taboo against women at sea, including comments she solicited from several staff members. The most notable of these came from physical oceanographer Henry Stommel, who said, "Part of the issue is an emotional one. For many men, going to sea represents a temporary reversion to boyhood, and they don't want their glorious chance to get away from it all endangered. I should think WHOI might be realistic about it and take a couple of week-long trips to see how it works. If the Institution could make a ship go out of here that was fit for women to be aboard, it would probably be a better research vessel for all concerned, even if no women ever went out on it."

Betty Bunce began working for Navy 7 in the summer of 1944, when she was teaching in a private school. She later joined the geophysics group full time and in 1959 was the first WHOI woman to go to sea on an Institution vessel. She also served as chief scientist aboard a *Bear* cruise that year. This picture was taken in the *Atlantis II* main lab in 1977.

When nothing had changed the following year, *Caryn* left the pier on a July cruise with Eike stowed away in the bilge. Upon discovery, she was confined alone to the captain's cabin, while some of the planned research was undertaken. Chief scientist Clarke administered a spanking, the cruise returned early, and Eike's fellowship was terminated.

Three years later, geophysicist Betty Bunce served as chief scientist aboard R/V *Bear,* and Eloise Soderland and Lorraine Barbour participated in a physical oceanography cruise aboard *Chain* in 1960. Slow but steady acceptance of women scientists at sea had begun; for crew members it was delayed until 1977, when Deborah Mennett sailed as corpsman and Martha Coneybear as ordinary seaman with *Alcoa Seaprobe* (which was on a trial loan to the Institution). The engine room barrier was broken by Kathleen Hoard aboard *Knorr* in 1981. The short list of thirty-four *Alvin*

Eloise Soderland, left, and Lorraine Barbour also helped lead the way for women at sea when they processed hydrographic data aboard *Chain* during Gulf Stream '60. Reporting on that program, *Oceanus* magazine said, "The girls worked some 10–14 hours a day to keep up with data radioed in from the other ships and those taken by the *Chain*."[118]

pilots includes one woman, Cindy Van Dover, who was fascinated by sub operations during her biological graduate work in the MIT/WHOI Joint Program and spent three years with the *Alvin* group following completion of her Ph.D. in 1989. She made forty-eight

dives as pilot-in-command (including one with two female scientists) before returning to science full time.[119]

In 2005, the crew and officer rosters often include at least one woman, including Mitzi Crane, who has made several trips as master of R/V *Atlantis.* The WHOI Women's Committee honored four Women Pioneers in Oceanography during the 1990s. At the end of the decade, the Institution formally established the Mary Sears Woman Pioneer in Oceanography Award, and it was first given in 2002. The recipients of these awards are listed in Appendix II, on page 165.

Sallye Davis handles a line during an *Atlantis* cruise in 2000. She first worked aboard a WHOI ship as a messperson, in 1984, then served in ordinary and able-bodied seaman positions for several years, becoming a third mate in 1991.

The *Crawford* Caper

With Bermuda as the last port of call, *Crawford* arrived at the WHOI pier the afternoon of March 19, 1960, and quickly cleared customs and immigration with the usual New Bedford agents. When unloading was nearly finished, four Boston customs agents materialized to confiscate the booze that, by long-standing practice, cruise participants had been bringing home. The source of the tip the Boston agents were apparently acting upon was never revealed. More than eighty cases were confiscated, two personal vehicles were seized, and *Crawford* captain David Casiles was charged with allowing illegal importation. *Chain,* due in three days later, was cryptically notified, and Captain Hiller put all the liquor aboard the ship in bond (for later seizure, as it turned out). *Crawford* was allowed to continue operating, although it was for a time officially the property of U.S. Customs (which assigned custody to Captain Casiles!). A three-person internal investigating committee determined that the liquor was not intended for commercial purposes, and payment of a $4,500 fine (contributed by those who had benefited from the contraband over the years) cleared Captain Casiles. New policies were put in place, but life with customs was difficult for some time.[120]

The Henry Bryant Bigelow Medal was established in 1960, and the first award went to its namesake in 1961. Henry Bigelow, left, and WHOI Director Paul Fye admire the medal following the medal presentation in Cambridge.

As the Institution celebrates its thirtieth anniversary in 1960, Henry Bigelow and Paul Fye jointly sign the director's message in the annual report, noting that the variety of work published in 1,134 numbered WHOI contributions to the scientific literature indicate that the Institution has "indeed carried out the mandate of our charter 'to prosecute the study of oceanography in *all* its branches.' " They also express pride in "the continued excellent cooperation between scientist and sailor and between master and chief scientist . . . demonstrated on every cruise." August

10 is declared "Bigelow Day." It features a luncheon honoring the first director, and Paul Fye announces establishment of the Henry Bryant Bigelow Medal in Oceanography. The first of these medals will be presented to Bigelow himself at a quiet ceremony at the Harvard Faculty Club in Cambridge on March 16, 1961. (See pages 164–165 for a list of the succeeding recipients.)

The publication date for the first edition of the WHOI *Newsletter* is April Fools' Day, 1960. The three-page document shows architect's drawings of a metal Butler building that will later be named Blake (for a U.S. Coast and Geodetic Survey steamer that made the first classic oceanographic study of the Gulf Stream in the late nineteenth century). The newsletter notes that the business offices are moving from the "white house" to Challenger House, where there are also still some overnight accommodations for visitors. "Ship News" reports that Dick Edwards is the newly appointed chair of the Ship Design Committee and describes "Gulf Stream '60," the year's major effort at sea: "the three ship synoptic survey of the Gulf Stream begins Monday with the departure of *Chain* followed in a day or so by *Crawford* and *Atlantis.* Work will begin immediately with *Chain* making sixty-seven stations, *Crawford,* sixty-two, and *Atlantis,* fifty-four. Each ship will operate independently, parallel to one another, on north to south tracks from the Grand Banks to Bermuda. The first phase of the survey will be led by Val Worthington subbing for Fritz Fuglister who has been in sick bay."

There is considerable interest among both the Navy and oceanographic communities in deep-sea exploration with human-occupied vehicles. In January 1961, Charles B. "Swede" Momsen Jr., the chief of undersea warfare at the Office of Naval Research, secures $1 million for a year's lease and operation by WHOI of *Aluminaut,* a 51-foot (15-meter) sub being built by Reynolds

The five-thousandth *Atlantis* hydrographic station was recorded in 1960. This photo shows Arnold Clarke making one of those stations. In March 1962, *Oceanus* magazine gave this description: "A hydrographic station is one of the basic operations in oceanography. To make a station, a series of Nansen bottles with attached thermometers are lowered into the ocean to obtain temperatures and water samples at various depths from surface to bottom. Completing a station takes several hours depending upon the depth to be sampled. Usually two 'casts' are made. One above 2000 meters with close-spaced bottles, and one from 2000 meters to the bottom with more widely-spaced bottles."

Frank Mather pulls a tuna aboard *Crawford* for game-fish studies.

Game-Fish Studies

Frank Mather ran a game-fish tagging program that was based at WHOI for twenty years before becoming a cooperative effort with the National Marine Fisheries Service. This summary of the program's first six years appeared in the June 1960 issue of *Oceanus* magazine:

Two bluefin tuna tagged and released in 1954 off No Mans Land, Mass. were recaptured in August 1959 by French fishermen in the Bay of Biscay. One tuna grew from an estimated 18 pounds to about 150 pounds in its five years of freedom. This exciting news was of great interest as most biologists hold to a theory that the eastern and western Atlantic bluefin populations are distinct races. Apparently, they do intermingle.

After years of tenacious work and patient waiting the efforts of hundreds of co-operating sports fishermen have started to bring results. Since 1954, when Frank J. Mather III of our staff started the game fish tagging program, some 766 tuna, 1000 white marlin, 475 Atlantic sailfish and 230 Pacific sailfish have been tagged, as well as smaller numbers of other gamefish. Obviously, biologists have neither time, nor the money to do much tagging, nor are many excellent anglers. Mr. Mather is. We are most grateful that hundreds of sportsmen have become interested in the program and have released so many fish. In some tournaments the organizers now award points and prizes for tagged and released fish.

The returns have shown that some original fears—that released fish might not survive after the struggle of capture and wounding by the tag—were groundless.

Metals Company. Just as that arrangement begins to fall apart (when Reynolds wants more money), Lieutenant Don Walsh of the Naval Electronics Laboratory in San Diego, which operates the cumbersome bathyscaphe *Trieste*, arrives in Momsen's office with a design for a small, maneuverable sub his group wants to operate under the name *Seapup*. To Walsh's chagrin, Momsen diverts *Seapup* to WHOI, where an operating group is already assembling. Bids are invited for a submersible "capable of diving with two men 6,000 feet into the

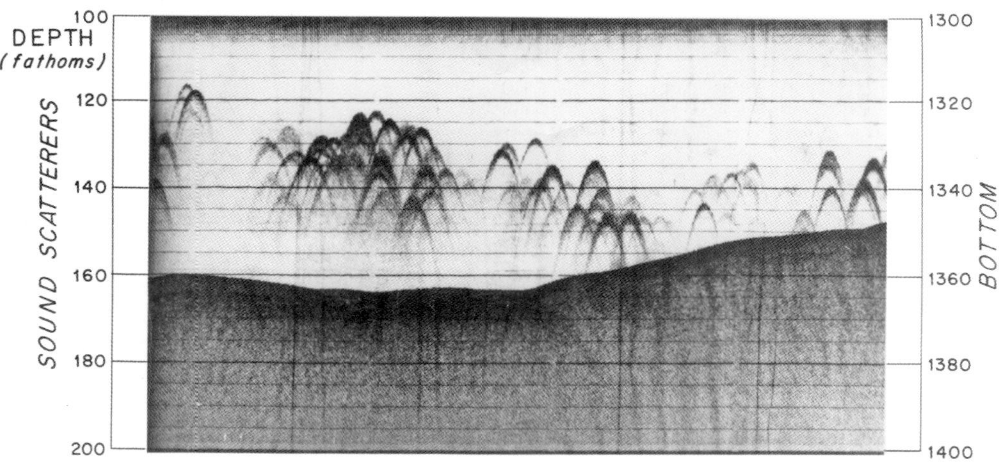

Acoustic traces like these were the object of many years' study by the geophysics group. In an early interdisciplinary move, group leader Brackett Hersey hired biologist Dick Backus for these investigations. The figure shows a recording (circa 1960) of "Alexander's Acres," one among many reflection patterns that plagued naval personnel looking for submarines. ("Alexander" was Lieutenant Sidney Alexander, skipper of the USCG Cutter *Yamacraw,* used by the geophysics group in 1957 and 1958.) Backus ultimately was able to positively identify "Alexander's" fish when he could see them from *Alvin* during a dive in 1967. They proved to be large schools of 2.5-inch (63-millimeter) myctophid or lantern fish, *Ceratoscopelus maderensis.*

ocean to make observations and collections never before possible." The contract goes to the Applied Science Mechanical Division of General Mills, where *Seapup*'s designer Bud Froehlich is located. The sub will become *Alvin,* named for Allyn Vine, who was instrumental in convincing both Navy and civilian communities that taking humans to the deep sea was both important and feasible. (*Alvin*'s milestones will be noted in this text, but readers are referred to Victoria Kaharl's *Water Baby: The Story of Alvin,* for the full and fascinating tale of the sub's first quarter-century.)[121]

A *Chain* cruise in late 1961 included a port call in Monaco, where Prince Rainier and American-born Princess Grace accepted Captain Hiller's invitation to Thanksgiving dinner. Here Gerry Metcalf welcomes the princess aboard. Prince Ranier, left, shepherds two royal children, and Captain Hiller, in uniform, is behind Princess Grace. Jacques Cousteau's wife, Simone, graciously assisted the ship's company with protocol for the occasion.

The First Seagoing Computer

Carl Bowin, who was hired just out of Princeton in 1961 by Brackett Hersey to start a marine gravity program, tells about the Institution's first seagoing computer:

I went down to Washington to see how the Coast and Geodetic Survey made its measurements, because it had started making sea measurements along the coasts. I remember seeing this room with five tables and women at these tables reading the records which came off their . . . instruments, which had a strip chart. It took them three days to process one day's data! With this system, you wouldn't have the results until months after you got back to shore.

It took me three milliseconds to think, I've got to put a computer together with the instrument, at sea, so we can see the results in real time. At that time, Brackett Hersey had lots of navy money, so he said, "Yes, we'll do it." In tack-

Working with his sea gravimeter aboard *Chain* in 1962, Carl Bowin makes the first WHOI use of a computer at sea. Bowin reported a "thrilling sense of importance" when the system's type-writer "printed the world's first values of gravity to be collected and reduced automatically at the same time as the measurements were made."[122]

ling this problem, I interviewed all of the major computer companies, and the one that seemed to have the most interest was IBM. But the engineer who designed the [processing unit, a "1620" office computer] said it wouldn't last [15] minutes at sea. . . . It was Brackett—I had never been to sea before—who said, "Sure. It will work!"

We rented the IBM machine (it was too expensive to buy) plus we paid monthly fees for an IBM maintenance guy to go to sea with us. It was very costly. Also the research vessel Chain *did not have air conditioning at that time. We had to build a special room within the main lab which was air conditioned for the computers. It was the very favorite room for people to visit at sea in the tropics, like off Puerto Rico.*[123]

By 1969, shipboard scientific computing and satellite navigation were becoming standard on WHOI cruises.[124]

Changing Times and the "Palace Revolution"

The appointment of Paul Fye as director in 1958 marked the beginning of WHOI's coming of age as an organization. The early days of "gentleman science" were long gone, and the labs and ships were inhabited by people driven to learn as much as possible as quickly as they could. They liked the Institution's free atmosphere, and they found Iselin to be an approachable, sympathetic director, whose "organization" involved sitting down with individual investigators to discuss what they needed and giving it to them, if at all possible.

However, the board of trustees determined that the time had come to create a more formal, businesslike structure. For the most part, the scientific staff didn't like the organization Fye was hired to bring to the maturing Institution. As he was finding his feet as director and spending considerable time in Washington, his choices for deputies were questioned. The staff felt slighted and ill-informed by a director who (though he didn't intend to be) appeared indecisive, aloof, and rather disinterested in the science. After careful consideration that included group meetings and memos on individual concerns, they decided to rebel.

Twenty members of the "palace revolution" group signed a petition to the Executive Committee asking for a chance to discuss the direction of the Institution and "the ineffectiveness of the administration." [125] While the Executive Committee gave them a hearing on Friday, December 19, 1961, it was soon clear that Fye's position was never in danger. Letters from board chair Noel McLean addressed to the "Resident Research Staff" arrived by mail as early as the following day, advising that "those who cannot in good conscience support the Institution should go elsewhere." [126] Eventually, some of the rebels did move to other institutions, while others found a way to live with the new regime and to share in a period of great success and expansion for the Woods Hole Oceanographic Institution.

1962–1965

By 1962, WHOI is moving full-speed ahead on all fronts. Navy funding totals some $4 million, about three-quarters of the Institution's total operating budget. There are two major events on June 18: a keel-laying for *Atlantis II* at the Maryland Shipbuilding & Drydock Company in Baltimore, and, with newly named associate director Buck Ketchum manning a bulldozer, a ground-breaking ceremony on Water Street for the Laboratory for Marine Sciences. (It will later be named for Alfred Redfield.) Both ship and construction are funded by NSF. The Institution acquires a DC-4 aircraft and also the 99-foot (30-meter) Army cargo vessel *Gosnold* for conversion to accommodate scientific work on day trips or short cruises.

Buck Ketchum, on the bulldozer, helped break ground for the Laboratory for Marine Sciences as one of his first duties as associate director for biology and chemistry. Others involved in the construction, from left, are Jim Mitchell, director's administrative assistant; David Scott, assistant director for administration; Antony Leonoe of Macomber construction; Redwood Wright, public information officer; and Patrick Walsh of Fuller construction.

Based on the recommendations of five visiting committees for the various disciplines, science departments are established for the first time "to place decision-making responsibilities closer to the scientists performing the research." The six departments (and their chairs) are Applied Oceanography (Earl Hays), Biology (John Ryther), Chemistry and Geology (Buck Ketchum), Geophysics (Brackett Hersey), Physical Oceanography (Fritz Fuglister), and Theoretical Oceanography and Meteorology (Columbus Iselin). Each chair contributes several pages to the 1962 annual report summarizing research in his department. The next step toward more formal academic organization comes in 1963 with development of formal appointment policies, including a new system of titles and tenure.

Brackett Hersey, left, and Paul Fye discuss WHOI research with visiting U.S. Navy sponsors in the 1960s.

Specifications for the Laboratory for Marine Sciences, later named for Alfred Redfield, included environmental rooms adjustable over a wide temperature range, air-conditioned rooms for special experimental and instrumental work, an aquarium, dissecting and X-ray facilities for ecological and icthyological studies, a computing center, and a 250-seat auditorium/lecture hall.

Atlantis II, just beginning its second research cruise in April 1962, answers a distress call from the Navy to join the search for USS *Thresher*, a nuclear submarine missing just east of Georges Bank. Battling repeated gales and navigation inade-

Fritz Hess transfers by highline from USS *Hazelwood* to *Atlantis II* during the *Thresher* search. Hess contributed his expertise with the survey equipment being used during the operation.

quate for the task, *A-II*, along with a cadre of WHOI people and instruments, spends nearly two months in a search that Brackett Hersey likened to "lowering a ping pong ball into a beer can from the top of a three story building while blindfolded and during a northeaster."[127] *Chain* also participates in late summer. Though WHOI and other research vessels photograph and dredge a few pieces of debris likely to indicate *Thresher*'s position, it is not until August that a dive in the Navy bathyscaphe *Trieste* fully confirms the nuclear sub's resting place. The *Thresher* search inspires better techniques for tracking deep-sea instruments, improved navigation systems, and Navy interest in small submersible operations.

Sydney "Bud" Knott, principal developer with Warren "Whitey" Witzell of the precision graphic recorder, a modern echosounder, monitors the machine during *Thresher* operations. Knott received the Navy's Meritorious Public Service Citation for his work during the search.

The new ship's next adventure, Cruise 8, beginning in July 1963, takes it to the other side of the world to participate in the International Indian Ocean Expedition, which also occupies WHOI's C-54Q aircraft for six weeks. Forty-one members of the WHOI staff join Cruise 8 scientific parties, and the chief scientists are Arthur "Rocky" Miller, Paul Fye, and Columbus Iselin.

 the optimum size for
the Institution through the 1960s. In 1963, Buck Ketchum writes, *For many
oceanographic problems a large institution is essential. The problems of operating
ships efficiently can be surmounted only by an institution large enough to support a staff for the planning and effective use of these vessels. The research staff
must be large enough so that each scientist has time to evaluate the results of his
cruises since no one can spend too much time at sea and still publish. There is no
simple formula, however, to relate ships, people, and shore laboratories. A geophysicist may reach the dock with much of his observations automatically
recorded, whereas a biologist may work months or even years on plankton collections before he reaches the same state of preparedness. It is therefore a difficult
question as to how large an institution of oceanography should become.*[128]

WHOI's Dean Bumpus initiated continental shelf–circulation studies in 1955 in conjunction with eight other institutions and government agencies. In 1962, the group released 23,000 drift bottles and 7,800 seabed drifters, recovering 10 and 15 percent of them, respectively. Resulting data contributed to preparation of twelve charts depicting direction and speed of the nontidal drift for the North American East Coast from Canada's Cabot Strait to Florida.

In addition, WHOI operates the 243-foot (74-meter) *Anton Bruun* (once
Truman's presidential yacht *Williamsburg,* but now renamed for a famous Dan-
ish oceanographer) for ten cruises from 1963 to 1965. These voyages support
NSF's Indian Ocean biological program, which is directed by John Ryther.
A-II's second trip to the Indian Ocean takes the ship on WHOI's first round-
the-world cruise, a voyage of nearly fifty thousand miles, from January to
November 1965. A six-month *Chain* cruise in 1964 also contributes geology and geophysics data to the UN-sponsored
international effort to explore the world's least-known ocean.

In February 1964, while steaming midway between Bermuda
and the Azores, *Atlantis II* encounters a forty-foot wave that
breaks several pilothouse windows, catapults the forward port-
hole from the chief scientist's cabin through a bulkhead into
the next cabin, and damages booms, frames, and electrical
equipment. Three people suffer cuts and bruises, and damage
to the ship totals $25,000.

The annual report terms the collection of iron-rich sediments
from the hot brine zones in the Red Sea a major discovery for
1965. Though previous bathymetric surveys have recorded high
temperature and salinity in the Red Sea rift trough, *Atlantis II*

Cruise 15 is the first to obtain cores of sediments beneath the hot, salty brines. This finding is expected to contribute to eventual solution of several geological problems, including the origin of heavy metal deposits and the influence of tectonism (crustal motion) on subsurface fluid movement.

WHOI enters two collaborative geological programs in the early 1960s, the first an alliance with the U.S. Geological Survey (USGS) to study the continental shelf and slope between Maine and Florida. The investigations begin in July 1962 with K. O. Emery as the principal WHOI participant working with four USGS scientists. By 1966, having employed 634 days of ship time, most of

In the photo at left below, a crowd of five hundred gathered on the WHOI waterfront in June 1964 for *Alvin's* commissioning ceremony. Allyn Vine's wife, Adelaide, was the sponsor, and speakers included the Honorable James H. Wakelin Jr., assistant secretary of the Navy for research and development, and Captain Charles B. Momsen, director of undersea programs for the Office of Naval Research. During a three-month operating period the first year, *Alvin* made seventy-seven local dives carrying fifty-four observers. At the end of the year, the sub was completely disassembled for inspection and minor refit before attempting deep dives in 1965.

At right, Howard Sanders's research group towed this newly developed deep-sea "epibenthic sled" horizontally to collect the uppermost part of seafloor sediment, yielding a new view of populations living there. Instead of being desertlike, the researchers found the seafloor rich in invertebrate fauna, "comparable to that of the shallow-water tropics."[129] The photo shows George Hampson, left, and Steve Page decanting collected sediment.

it aboard *Gosnold* and some for *Alvin* dives, the considerably enlarged group completes the initial studies and is ready to begin expanding its reach to the continental rise. In 1974, the USGS contingent will become the USGS Branch of Atlantic Marine Geology, which in 2005 numbers one hundred people occupying two buildings on the WHOI Quissett Campus.

The second geological program begins in May 1964 with an agreement among the oceanographic laboratories at Woods Hole, Scripps, Lamont, and Miami to establish the Joint Oceanographic Institutions for Deep Earth Sampling (JOIDES) program. Its ambitious purpose is to obtain core samples of the entire column of sediment on the ocean floor. The following year, JOIDES takes advantage of an offer from the Pan American Petroleum Corporation to use the company's drilling ship *Caldrill* to secure the program's first six cores from the Blake Plateau off the Atlantic Ocean side of Florida. (In the late 1960s, Woods Hole scientists will share the excitement with colleagues from other institutions when cores taken by the program's first ship, *Glomar Challenger,* convincingly demonstrate the existence of seafloor spreading.) Forty years and four operational programs later, deep-ocean drilling for long cores continues to play an important role in oceanographic research for scientists at Woods Hole and many other laboratories in the United States and abroad. Paul Fye was the first chair of the JOIDES executive committee, and succeeding WHOI directors and many members of the scientific staff, as well as the USGS Woods Hole branch, have been integral to the deep-sea drilling programs.

The Little Sub That Could

The U.S. Navy called *Alvin* to military duty to help find a hydrogen bomb accidentally dropped into the Mediterranean Sea near Palomares, Spain, in January 1966. Flown to Spain aboard an Air Force cargo plane, the sub operated from a Navy dock landing ship (LSD), designed to transport and launch amphibious craft and vehicles. *Alvin* and its

crew made thirty-four dives from February to April, first locating the bomb on March 15 and again on April 2 after it skidded down a slope on the first recovery attempt. The sub returned home a hero, only to be attacked the following summer during Dive 202 by an aggressive swordfish (which did little damage but made a fine dinner) and to lose its mechanical arm in the fall during Dive 224.

The arm was recovered three weeks later, but this accident foretold a much more serious one in which the sub itself sank when two lift-platform cables broke during a launch in October 1968. The three people inside (pilot Ed Bland, pilot-in-training Roger Weaver, and Paul Stimson of the Buoy Group) scrambled to safety, leaving their lunches behind. While there was some sentiment in the Navy for simply leaving *Alvin* at 5,000 feet (1,515 meters), recovery advocates prevailed, and the sub was raised ten months later along with its well-preserved sandwiches—which inspired a new line of deep-sea microbiological investigations initiated by Holger Jannasch.

During the search for a U.S. hydrogen bomb lost in the Mediterranean off Spain in 1966, *Alvin* operated from a Navy dock landing ship. The red beret was adopted as a badge of camaraderie by the *Alvin* group during its early years after pilot Val Wilson, originally hired as skipper for *Lulu,* came back from a trip to the Bahamas sporting one.

On October 16, 1968, two steel cables supporting *Alvin*'s lowering cage snapped at the beginning of Dive 308 one hundred twenty miles south of Cape Cod. The sub plunged about 15 feet (4.5 meters), then bobbed to the surface where it stayed long enough for pilot Ed Bland, scientific observer Paul Stimson, and pilot-in-training Roger Weaver to scramble to safety. *Alvin* then sank to the 5,000-foot (1,515-meter) bottom where it rested until recovery the following year. The photo shows the sub being lifted aboard a barge after its recovery.

1966–1969

On November 9, 1966, *Atlantis* departs Woods Hole harbor for the last time to serve under the flag of Argentina as *El Austral*. Henry Stommel captured the local sentiment with these words: "When the *Atlantis* was donated to Argentina, and she sailed away forever, something very fine was lost to Woods Hole. Not only a link, which bound everyone who had sailed in her, but a way of life at sea had vanished with her. There is nothing quite like a sailing ship at sea, the wind on the beam, in the moonlight the vast white sail curling upwards a hundred feet, the stays taut like great bowstrings, the balanced tensions, the full view fore and aft over the low deck houses, the exhilaration of the steady heel, the blessed silence without the rumble of engines, the bite of the working wind. And a good captain and a good crew."[131]

In January 1967, the first staff council retreat focuses on planning for a formal graduate education program and reorganizing the

Atlantis's service to U.S. oceanography ended in 1963. The ship was renamed *El Austral* and transferred to the Argentine Council of Scientific and Technical Investigations at this well-attended pierside ceremony in Woods Hole on November 9, 1966.

In March 1968, the seven members of the directorate and five department chairs, along with distinguished guests, made their second "retreat" at White Cliffs in Plymouth, Massachusetts, to discuss scientific and educational goals for the Institution. The participants, left to right, were Paul Fye (director), Burr Steinbach (soon-to-be dean), Roger Revelle (Scripps), John Ryther (biology chair), John Hunt (chemistry and geology chair), Nick Fofonoff (physical oceanography chair), Buck Ketchum (associate director), Scott Daubin (ocean engineering chair), Wilbert Chapman (a California seafood expert), K. O. Emery (acting dean), Fred Mangelsdorf (assistant director for development and information), Art Maxwell (associate director), David Scott (assistant director for administration), and Earl Hays (applied oceanography chair).

science department structure. The group decides that the formal education program should not have a separate faculty but rather should be integrated with the research program, which will place more responsibility on the department chairs. The new organizational structure for the scientific disciplines includes Departments of Biology, Chemistry, Geology and Geophysics, and Physical Oceanography (incorporating theoretical oceanography, meteorology, and the Buoy Project), as well as a new Ocean Engineering Department (in place of the Applied Oceanography Department).

By the end of 1967, the WHOI personnel count reaches 598 (including 318 in the science departments). Growth dictates land acquisition during the 1960s, and village properties are carefully evaluated when they come onto the market and purchased if they show promise for Institution purposes. The properties acquired are generally small lots, and many of the buildings are residences, which can be used for housing or converted to offices. Preliminary plans are drawn for a large laboratory on the Fay property in Woods Hole, but they are abandoned because of the crowded conditions in the village. Clearly, space for another large laboratory as well as future expansion is needed, and the trustees have, in fact, been watching since the mid-1950s for a large tract to purchase.

They especially monitor the 139-acre Fenno property a mile and a half from Water Street, with frontage on Woods Hole Road and Vineyard Sound. When it becomes available, there are two bids, one from the Institution and the other from a developer, who manages to delay the sale for some time. In October 1968, the title is at last transferred on the $475,000 property, and it is designated the "Quissett Campus." The following year, a five-year lease is negotiated with

Located on an enlarged pier, the Marine Services Building (later named for Columbus Iselin) opened for business in fall 1969, accommodating marine offices, the acoustics and electronics groups, the mechanical and electrical shops, and staging areas for cruises. The new 40,000-square-foot (3,600-square-meter) pier provided 740 linear feet (224 meters) of large-vessel berthing space, while the smaller boats moved to the Eel Pond channelside. These facilities were funded by the National Science Foundation.

the National Academy of Sciences for use of the residence and the carriage house on the property to house summer study groups. The 1970 purchase of the Webster estate, adjacent to the Fenno property on its northern side, will expand the Quissett Campus by 40 acres (and acquisition of the Bell property, to the south, in 1983 will add another 6.5 acres). The first construction on the new campus provides a small wooden-frame laboratory on the waterfront for studies of chemical "messengers" in marine animals.

With gifts from the Atlantic Foundation and J. Seward Johnson, a distant property, 50 acres of nearly pristine land where the Matamek River flows into the St. Lawrence, is purchased in 1966 in Quebec. Its few fishing-camp cabins will house ecological studies of an unpolluted estuary, including work on the Atlantic salmon's life cycle, in conjunction with the University of Waterloo, Ontario. (After more than twenty years of research there, the property will be sold in late 1988.)

Knorr made a big splash at Defoe Shipbuilding Company in Bay City, Michigan, in August 1968. *Knorr* and sister ship *Melville,* assigned to Scripps, were funded by the U.S. Navy.

R/V *Knorr,* being built by the U.S. Navy and named for a nineteenth-century Navy cartographer, is launched at Defoe Shipbuilding Company, Bay City, Michigan, on August 21, 1968. It will be delivered to WHOI on April 15, 1970. *Crawford* retires from the fleet in 1968 and is sold to the University of Puerto Rico in 1970.

An "Unprecedented" Joint Education Program

When WHOI was very young, professors conducted summer work alongside their students both in the laboratory and aboard *Atlantis* and *Asterias.* There were fellowships available to students from the very beginning. Later, as university interest in ocean science grew, there was concern that "young blood" would be drawn to those academic programs and that WHOI might become just a ship station.

Norris Rakestraw, dean at the Scripps Institution of Oceanography, and formerly an early WHOI chemistry associate while on the faculty of Brown University, was invited to spend the summer of 1963 reviewing WHOI's educational activities. He recommended several steps the Institution might take to strengthen its role in education, including appointing a director of education and establishing a student center. "If the Institution is to be an equal partner with the universities in the graduate training of students in oceanography," he wrote, "it must do more than merely provide facilities; it must actively participate in the academic process." While Rakestraw did not recommend the establishment of a degree program, the report raised questions that pointed in that direction.

At the 1964 annual meeting of members and trustees, Paul Fye suggested the possibility of giving "accredited courses at Woods Hole" and awarding "a joint degree in collaboration with selected universities." The trustees appointed a special Education Committee* at that meeting and charged it with determining whether and how the Institution could conduct graduate education without compromising its traditional commitment to research. After considering various approaches over the next two years, the committee recommended establishment of joint degree programs with both MIT and Harvard. Though some Harvard faculty members were vehemently against the idea and not all the relevant faculty at MIT were enthusiastic, a joint agreement was signed with MIT in 1968 (a move called "unusual, if not unprecedented," by the journal *Science*[132]). Some WHOI staff members were also less than enthusiastic (and a few were adamantly opposed), but as time went on and the program

Aboard R/V *Chain,* MIT president Howard W. Johnson signs the historic memorandum of agreement creating the MIT/WHOI Joint Program in Oceanography on May 8, 1968. WHOI director Paul Fye is seated at right, and the other witnesses are, from left, WHOI dean of graduate studies Burr Steinbach, MIT provost Jerome Wiesner, and Frank Press, head of MIT's Earth and Planetary Sciences Department and MIT director of the Joint Program during its first ten years of operation.

---

* The committee's initial members, James S. Coles (Bowdoin College), Arnold B. Arons (Amherst College), and Carroll L. Wilson (MIT), were joined the following year by Detlev Bronk (the Rockefeller Institute, which became The Rockefeller University in 1965) and Jerome Wiesner (MIT).

thrived, degree-seeking students were gradually woven into the Institution's fabric. Cross-registration arrangements were eventually made with Brown, Harvard, and Yale.

While MIT and WHOI each retained autonomy in matters relating to the separate institutions, the program was indeed a joint venture. A. Lawrence "Jake" Peirson III, who was associated with the program from its beginning (he retired in 1996 as WHOI associate dean and registrar), described the working relationship this way: "The truly unique thing about this program is that every decision—from the time students are

Four Joint Program degrees were conferred at the first formal graduation ceremony on the WHOI pier in 1970. The festivities were witnessed not only by family and friends but also by the new marine facility building and the just-delivered research vessel *Knorr*. Julius Stratton delivered the commencement address. Formal graduation ceremonies have been held in Woods Hole roughly every decade since then.

admitted to the time they get their degrees—is made by joint faculty committees whose members have to come to a consensus."[133]

The program was up and running quickly. Fifty-one students were enrolled in 1969, and four degrees were awarded in 1970. By the tenth anniversary, seventy-nine graduate degrees had been awarded, and the second decade began with eighty-nine students. In June 2004, program enrollment stood at 139, and degrees conferred to date numbered 145 master's, 57 engineer's, and 486 doctoral degrees. In addition, four doctoral degrees had been awarded by WHOI alone.

Generous gifts of $8 million from J. Seward Johnson and $5.5 million from Mr. and Mrs. W. Van Alan Clark launched the Joint Program, and private gifts have continued to provide vital support since then. "We take responsibility for supporting our students for the full five or six years it takes to earn their degrees," said vice president for academic programs and dean John Farrington. Rather than having to pursue another scientist's already financed research, each Joint Program student is free to choose an original thesis topic, find the right advisor, and make adjustments as needed.[134]

While focusing the traditions, practices, and opinions of two institutions on one Joint Program has not always been easy, it has been successful. Robert Morse, WHOI dean from 1973 to 1979, perhaps put it best when he said, "I don't think any other graduate oceanography program got students of such high quality, and it isn't clear that either MIT or WHOI could have gotten those students themselves."[135]

Each summer, students entering the Joint Program participate in a week-long group adventure aboard a Sea Education Association ship. Here students and members of the brigantine *Westward*'s crew haul in a line at the beginning of the first "oceanographic orientation cruise" in 1990. Fifteen new students and eight upperclass students participated that year.

The fellowships awarded to students beginning in 1931 continued over the years in various formats. The Summer Student Fellow program was formalized in 1959 to focus on upper-level undergraduate and beginning graduate students, and shortly thereafter a more formal program for supporting postdoctoral fellowships was also instituted. Both programs are highly competitive, attract outstanding applicants, and are listed on the résumés of many WHOI staff members.

Release in January 1969 of the influential report *Our Nation and the Sea,* better
known as the Stratton Commission
Report,* assigns, Paul Fye writes, "new
national perspective to the young science of oceanography. . . . [It] may well
be the most important single document concerning the oceans in our lifetime." It charts a course, he continues,
that "can insure that the oceans will
benefit all mankind for generations to come through the leadership of the
United States." The commission offers 126 recommendations ranging from
details for organizing the national oceanic effort to U.S. government reorganization for greater efficiency and productivity.

Two of the recommendations most likely to affect oceanographic institutions are (1) the setting up of a new agency to oversee the government's civilian
activities in the oceans and (2) the establishment of "three or four major
University-National Laboratories." Both recommendations cause concern that
the independence of ocean science laboratories may be compromised; however,

the report also includes an important
endorsement for basic science: "Continuing and substantial support of
basic marine science is a national
investment which will provide an
underpinning for all future activities
in the sea."[136] WHOI will form a good
working relationship with the
National Oceanic and Atmospheric
Administration, established in 1970,
and Paul Fye will be a leader in bal-
ancing the power of federal agencies
and individual institutions as the University-National Oceanographic Labo-

---

* The commission's proper name was the Commission on Marine Science, Engineering and Resources, but it
was generally known as the Stratton Commission for its chair, Julius Stratton, chairman of the Ford Foundation and former president of MIT.

ratory System is organized to bring greater efficiency to the use of ships and other U.S. marine facilities. A third Stratton Commission recommendation, endorsement of the International Decade of Ocean Exploration, will nearly double NSF's support for ocean science in the 1970s.[138]

1970–1974

A week of fortieth-anniversary festivities in June 1970 includes a two-day public open house, the first commencement of the MIT/WHOI Joint Program with four graduates and Julius Stratton as the speaker, and dedication of the Columbus O'Donnell Iselin Marine Facility (consisting of a new building, pier, and courtyard). There is also a dedication of the original WHOI building as the Bigelow Laboratory, awarding of the fifth Bigelow Medal to Princeton professor Frederick John Vine, and an address by William D. McElroy, director of the National Science Foundation.

Increasing societal concern about the environment is reflected in the frequent appearance of the phrase "to use the oceans wisely" in WHOI reports and discussions. In 1970, WHOI biologists along with Boston University Marine Biologi-

Vice President Hubert Humphrey examines several seabed drifters during a cruise he made aboard *Atlantis II* in 1967.

In 1970, John Teal and colleagues initiated a study of salt marsh response to such disturbances as sewage pollution. In the photo, Teal describes the project to a group of students.

cal Program colleagues, who are based at MBL, begin a series of experiments on nearby Great Sippewissett Salt Marsh. Their objective is to predict how disturbances of the environment (such as sewage pollution) can affect plant growth on the marsh and export of organic material to the surrounding water. The same year WHOI scientists place tanks on the pier for a pilot aquaculture experiment. It is designed to investigate the feasibility of using sewage effluent to nourish phytoplankton that will be fed to shellfish in a system intended to produce clean water and a commercially valuable product.

In the 1971 director's report, entitled "Oceanography and the Environment," Paul Fye calls attention to recent congressional testimony predicting "utter catastrophe for the marine environment before the end of the century." In response, he poses pertinent questions: "Are oceanographers capable of pointing the way to prevention of marine disaster, and are we making our full contribution? . . . In the cycle of life in the sea, who can say how delicate the balance is?" He also laments the handicap imposed on ocean scientists "because a suitable base of knowledge was never adequately established by scientists a hundred years or more ago before man began to pollute the oceans." Fye summarizes WHOI's current and past contribution to marine pollution studies, including the surprising finding that, because pollutants are as abundant in the open ocean as in coastal waters, they must travel not only by river, as expected, but also through the atmosphere. He calls oil pollution "one of today's most alarming pollution possibilities" and describes Institution scientists' efforts to monitor the long-term consequences of a 1969 barge spill in Buzzards Bay near

In the mid-1970s, Howard Sanders collects samples in the intertidal area of Wild Harbor in West Falmouth, Massachusetts, where the aftereffects of a 1969 oil spill have been monitored by WHOI scientists for more than thirty-five years.

West Falmouth, Massachusetts. (The spill will still be evident in Wild Harbor sediments as the monitoring continues thirty-five years later.)

The WHOI Marine Policy and Ocean Management Program, initiated in 1971 with Ford Foundation funding, is based on the premise that "Wise use of the oceans depends on more than wisely planned research. It depends upon the capability, willingness and wisdom of policy makers around the world." [139] The program, which broadens the Institution's mission beyond scientific research, is intended to contribute to international cooperation, provide scientific input to policy formation, and help ensure that the oceans remain free for scientific research (need for the latter is underscored by "a long series of legal and bureaucratic harassment" regarding work in the Red Sea during *Chain*'s 1970–71 circumnavigation). J. Seward Johnson provides three postdoctoral fellowships for the program. Over the years marine policy staff, fellows, and visitors will consider topics ranging from coastal zone and fisheries management to global warming, biological diversity, offshore energy resources, and the economic impacts of non-native aquatic nuisance species. In 2005, the objective of the Marine Policy Center is to conduct social scientific research that advances the conservation and management of marine and coastal resources by integrating economics, policy analysis, and law with the Institution's basic research in ocean sciences.

WHOI is selected as the first institution to host and operate the University-National Oceanographic Laboratory System (UNOLS) under a grant from NSF, the Office of Naval Research, the National Oceanic and Atmospheric Administration, and the Environmental Protection Agency. WHOI provost Art Maxwell is the first UNOLS chair, and Robertson P. Dinsmore, whose first Woods Hole assignment was with the Coast Guard's Ice Patrol in the early 1950s, is hired as executive secretary. UNOLS members include seventeen major U.S. laboratories, which operate thirty-two seagoing research vessels. UNOLS coordinates ship schedules and seeks seagoing opportunities for scientists who would otherwise not have direct access to ships.

Planning is under way in 1971 for new facilities ashore and afloat. These include WHOI's largest building (80,000 square feet, or 7,400 square meters, in three

The top photo shows *Knorr* during a port call in New York City at the end of a 1973 cruise to host a reception for members and representatives of the United Nations Seabed Committee. More than six hundred attended.

In the lower photograph, *Asterias* welcomes *Chain* home from a twenty-month (1970–71) around-the-world cruise that took investigators to the North and South Atlantic, the Indian Ocean, the Somali Current, the Red Sea, the Java Sea, across the Pacific, through the Panama Canal, into the Caribbean, and back home to Woods Hole.

The oceans have become a political issue. . . .
The concept of freedom of the seas is increasingly under attack. . . . the developed nations of the world must find better ways of assisting the developing world to make use of the oceans about them.[140]

—Paul Fye, 1972

When the 33-year-old bascule bridge over Eel Pond channel was being repaired in 1971, Al Vine and colleagues rigged this tire ferry to keep foot traffic moving on Water Street in Woods Hole.

Director Paul Fye, left, and Ken Leach of the Boston construction firm George B. H. Macomber break ground in March 1971 as construction of new buildings begins on the Quissett Campus.

connected structures on the Quissett Campus) and a second smaller laboratory to house aquaculture research. A $50,000 grant from the Fleischman Foundation supports design of a new ship (which will be *Oceanus*). The Institution's collection of bottom samples, cores, and seafloor data moves to a new 12,000-square-foot (1,115-square-meter) building near the Quissett Campus entrance in November 1971.

In July 1972, *Knorr* begins Atlantic sampling for the Geochemical Ocean Sections Study (GEOSECS), which Chemistry Department chair John Hunt describes as "the most ambitious project ever proposed to obtain detailed hydrographic and chemical measurements in the world's oceans."[141] *Knorr*'s ten-month GEOSECS odyssey from north of the Arctic Circle to the fringes of the Antarctic precedes other ships' work for the program in the Pacific and Indian Oceans. An IDOE program, GEOSECS involves nineteen institutions and includes scientists from Japan, India, Italy, France, Germany, Canada, and the United States. A small building is constructed on the Quissett Campus to house the program's "water library," roughly fifty 30-liter samples from each of 120 stations in the Atlantic alone, as well as some larger volume samples. Derek Spencer, senior

Sister ships *Knorr* (WHOI) and *Melville* (Scripps) traveled the lengths of the Atlantic and Pacific oceans, respectively, in 1973–74 for the international Geochemical Ocean Sections (GEOSECS) study. WHOI's Derek Spencer, second from right, was a member of the GEOSECS executive committee and chief scientist on the *Melville* cruise pictured. Arnold Bainbridge, at right, of Scripps was the program's technical coordinator, and the rosette water sampler at left was one of its principal tools. A warehouse, still known as "GEOSECS," was built on the Quissett Campus to house water samples collected during the program.

scientist in the Chemistry Department, is a member of the three-scientist GEOSECS executive committee and is chief scientist for the first four cruises. The major objective of the program is to improve understanding of the origin and circulation of deep water in the oceans by providing a wide variety of systematically collected baseline data on ocean chemistry. GEOSECS takes advantage of new, highly maneuverable ships (*Knorr* and sister ship *Melville,* operated by Scripps) as well as advanced technologies that allow detection of trace substances contained in seawater. These technologies include mass spectrometers, low-level radioactivity counters, and gas and liquid chromatographs.

Mooring flotation goes over *Chain's* fantail during the Mid-Ocean Dynamics Experiment (MODE) in 1973. Though it was conceived a decade earlier, this investigation of deep-water dynamics had to wait for improvements in instruments and mooring technology. An early big-science project under the International Decade of Oceanography umbrella, MODE involved fifteen institutions and five research vessels. Later in the decade, the follow-on POLYMODE program had a wider geographical reach and included Russian investigators. Part of its name was drawn from a 1970 Russian experiment called POLYGON.

The Environmental Systems Laboratory was built to house aquaculture research at a time when there was considerable interest in how much of a hungry world's food supply might come from the sea. The ponds in this view of the lab were designed to contain treated and diluted sewage effluent for growing algae that, in turn, fed shell- and finfish residing in raceways located to the right of the building. The facility, designed by Kramer, Chin & Mayo of Seattle to filter and heat 1,000 gallons (3,800 liters) of seawater per minute, was the first large-scale operation to draw on salt water and treated effluent for aquaculture. John Ryther was the leader of this project.

The aquaculture project moves from the pier to the new Environmental Systems Laboratory on the Quissett Campus in 1973. Winner of a national merit award for engineering excellence, the 4,500-square-foot (405-square-meter) building features an elaborate complex of pipes and tanks for heating, cooling, and filtering seawater, and there are outdoor algae ponds and "raceways" for growing shellfish.

The largest move in the Institution's history takes place in 1974 when some members of each science department relocate, along with the directorate, to the new central laboratory on the Quissett Campus (see photo on pages 104–105). The director writes in the 1974 annual report that the new building, designed by Walker O. Cain and Associates, increases assignable laboratory and office space in major Institution buildings by 70 percent and "has done much to meet our critical needs for research and teaching facilities." At its October 1974 dedication, the building is named Clark Laboratory in honor of major WHOI donors Edna McConnell Clark and W. Van Alan Clark. Additional funding for the laboratory comes from The Atlantic Foundation, Robert Sterling Clark Foundation, The Arthur Vining Davis Foundations, The Henry L. and Grace Doherty Charitable Fund, Inc., Charles Hayden Foundation, The Kresge Foundation, The Andrew W. Mellon Foundation, Richard King Mellon Foundation, Pew Memorial Trust, and Surdna Foundation, Inc.

From left, Paul Fye, W. Van Alan Clark, and Edna McConnell Clark return to their places on the podium following dedication of the new central laboratory on the Quissett Campus as the Clark Laboratory in October 1974.

In 2005, the Clark Laboratory and its 1988 addition, Clark South, house the departments of Geology and Geophysics, Marine Chemistry and Geochemistry, and Physical Oceanography as well as the Academic Programs office. The other two science departments are based in Woods Hole village, Biology in the Redfield Laboratory and Applied Ocean Physics and Engineering in the Bigelow and Smith laboratories.

Knorr's crane passes a towline to *Lulu* as the two ships and *Alvin* (secured to *Knorr*'s deck) head to the Mid-Atlantic Ridge for the French-American Mid-Ocean Undersea Study in 1974.

An ambitious, three-submersible diving program culminates the four-year French-American Mid-Ocean Undersea Study (Project FAMOUS) from June to September 1974. This program, which includes Canadian and British scientists, gives investigators their first opportunity to examine in person the plate tectonic processes at work on the Mid-Atlantic Ridge, part of the 36,000-mile (60,000-kilometer), globe-circling mid-ocean ridge. *Alvin* is equipped with a new titanium sphere that nearly doubles the sub's depth rating from 6,000 feet (1,818 meters) to 10,000 feet (3,030 meters). Armed with mapping, photography, sampling, and other data from twenty-five prior cruises in the target area, the international team of scientists carefully plans some forty dives for *Alvin* and the French bathyscaphe *Archimède* and submersible *Cyana*. Despite a highly successful program, continued support for *Alvin* and *Lulu* appears tenuous in the fall of 1974 until three agencies agree to a three-year program of joint support for the sub as a national facility under UNOLS. The National Science Foundation, Office of Naval Research, and National Oceanic and Atmospheric Administration alliance to support *Alvin* will continue into the new millennium.

The WHOI newsletter reported that *Oceanus* was greeted on its November 21, 1975, arrival in Woods Hole, "with every whistle in town, a cheering crowd, and a brave Coast Guardsman providing the traditional spray from the chilly bow of a 41-footer." Painting and other final outfitting for the 177-foot (54-meter) ship were completed later. *Oceanus* was funded by the National Science Foundation and constructed by Peterson Builders, Inc., of Sturgeon Bay, Wisconsin.

1975–1977

R/V *Oceanus* arrives in Woods Hole in November 1975, and *Chain* retires from the fleet at the end of its 129th cruise in December, having logged 600,000 miles (960,000 kilometers) for ocean science. The following

year, research at sea takes *Knorr, Alvin,* and *Lulu* to the Cayman Trough in the Caribbean Sea and *Atlantis II* to the Indian Ocean and the western Pacific (Banda Sea). *Oceanus* gains its sea legs beginning in April 1976 with a varied schedule that includes mooring recovery, taking two hundred WHOI Corporation members and associates to watch the Tall Ships parade out of Narragansett Bay, and, later in the year, two blustery cruises to the site of the *Argo Merchant* tanker oil spill off Nantucket. *Atlantis II* spends the year-end holidays at sea, steaming from Singapore to India.

WHOI researchers say goodbye to computer cards in 1975 as WHOI's Sigma-7 is converted from single-stream batch processing to multijob time-sharing. It can serve sixteen users at a time, and departments are expected to install their own terminals. Connection of the Sigma-7 to a national network of computers is in the planning stages. WHOI scientists work at the forefront of computer applications to scientific problems, and a VAX 11/780 (costing $250,000) will replace the Sigma-7 in 1978. Six years after that, there will be more than one hundred microcomputers at WHOI, and Information Services will predict that one day there may be a computer in every Institution lab. (Indeed, by 2005, WHOI computers will outnumber employees about 2.5 to 1, and everyone will be networked to the world through the Internet and Internet 2, designed to facilitate data exchange.)

Fye calls for enhancing applied research, ocean engineering, and marine policy in his 1975 annual report. During that year, Robert Frosch, long involved in U.S. naval and civilian oceanography, but most recently assistant executive director of the U.N. Environment Programme, is appointed associate director for applied oceanography. Two years later he becomes the administrator of the National Aeronautics and Space Administration; subsequently, the names of four oceanographic vessels are chosen for the space shuttle fleet: *Atlantis, Challenger, Columbia,* and *Discovery.* On its first mission, the shuttle *Atlantis* will carry a piece of the *A*-boat's mast into space.

In February 1977, a team of geologists heads to the Pacific's Galápagos Rift for work aboard *Knorr* and *Lulu* to make a detailed study of an area where warm water and dead clams were detected by a Scripps expedition the year before. Following placement of navigation transponders, the work begins with a 12-hour, 3,000-frame camera sled tow. When the film is ready to view, 13 of

the 3,000 frames reveal a striking phenomenon: a lava flow, 8,250 feet (2,500
meters) deep, covered with hundreds of
thriving clams and mussels in an area of
elevated water temperature. On the
cruise's first *Alvin* dive to investigate this
heavily populated area, pilot Jack Donnelly and scientists Bruce Corliss (Oregon
State University) and Jerry Van Andel
(Stanford University) enter a new world.

Scientists had their first close-up look at
hydrothermal vents in 1977 during *Alvin*
dives to the Galápagos Rift. Jack Donnelly, pilot on the first vent dive, took this
photo of unusual tubeworms native to
vent sites during a return to Galápagos
vents in 1979.

They find not only shimmering water that is visible evidence of the warm water
vents they seek but also unpredicted communities of exotic animals thriving
around them. Two years later, deep-sea biologists will visit the Galápagos vent
communities, and their geologist colleagues will identify extremely hot vents bil-
lowing "black smoke" from the seafloor on the East Pacific Rise. The vents and
other mid-ocean ridge processes will become a major focus of earth science
research and are still a hot topic a quarter of a century later.

After nineteen years as director, Paul Fye steps down in October 1977, and John
Steele arrives from Scotland to succeed him.

Paul M. Fye, Fourth Director

The ocean sciences and the Woods Hole Oceanographic Institution were ready for major
expansion in the 1960s, and Paul Fye was the right person to lead the boom. A physical
chemist (B.S. from Albright College in 1935 and Ph.D. from Columbia University in
1939), he taught briefly at Hofstra College. He left there in 1941 to become a research
associate for a National Defense Research Council project in underwater explosives, working with George Kistiakowsky, who would soon join the Manhattan Project.

The following year, Fye moved to Woods Hole to work with the underwater explosives
group at WHOI, where he applied his expertise in chemical additives, depth charge
behavior, and the destructive effects of underwater explosives. By 1946, he was the
group's research director. Following a brief appointment at the University of Tennessee in
1948, he moved to the Naval Ordnance Laboratory, where he managed explosives
research until the WHOI board called him to become director of the Institution in 1958.

Paul Fye and *Chain* (behind Fye) both brought profound change to oceanography in Woods Hole.

WHOI celebrated the fourth director's nineteen years at the helm with Ruth and Paul Fye Day in August 1977. Here the couple enjoys one of many presentations at the outdoor festivities held on the Quissett Campus.

Paul Fye brought management expertise, experience with government bureaucracy, and—probably most important—his vision to the Woods Hole position. In 1959, when the National Academy Committee on Oceanography recommended doubling basic ocean science research,[144] the Institution's Board of Trustees recognized "an expansion of the Institution to be a necessary part of its responsibility in the field of oceanography."[145] Fye quickly began implementing this directive, beginning with appointment of committees to consider scientific policies, new facilities ashore and afloat, and education programs. The board named him president of the Corporation in addition to director in 1961.

The 1960s were a time of astonishing growth for WHOI, and the 1970s, though less expansive, brought many new challenges. The number of employees doubled during Fye's tenure as director, and the Institution's operating budget increased nearly tenfold, from about $2.5 million in 1958 to $24 million in 1977. Bigger, more capable research vessels traveled the world, and a research submersible joined the fleet. New buildings accommodated the greatly enlarged staff, first in Woods Hole village and then on the new Quissett Campus. Fye worked tirelessly to initiate first a formal graduate education program and then a marine policy program (which he personally directed for eight years) and to secure the private funding that would ensure their success. He was a guiding light nationally as the maturing science of oceanography demanded inter-institutional cooperation.

Though he experienced some conflicts with staff early in his directorship (see page 115 on the "palace revolution"), Fye later encouraged the involvement of scientific and other Institution staff in discussions about growth, organization, and scientific direction. The elected Staff Committee was initiated in 1970 to "consider matters of institution policy and to advise the administration." The Women's Committee was added in 1973, and the Graded and Marine Personnel Committee in 1974.

Near the end of his nineteen-year term, more than seven hundred employees and friends gathered to honor the director and his gracious wife on Ruth and Paul Fye Day in August 1977. Paul Fye's guidance continued to be felt directly for many years (he served as president of the Corporation until 1986), and his vision is still apparent today in a robust institution built on sound principles instilled during his time as director.

1978–1979

Ferris Webster, associate director for research since 1973, becomes assistant administrator for research and development at NOAA, and Chemistry Department chair Derek Spencer takes the position of associate director for research.

As a cost-saving measure, *Atlantis II* is converted from steam to diesel power. This will also increase the speed and range of the vessel. Funding for the $1.8 million refit comes from NSF and WHOI, and ONR furnishes the new engines.

New microscope techniques allow the first identification of small, unicellular, marine blue-green algae thought to be responsible for 10 percent or more of the world oceans' primary productivity. First observed in the Arabian Sea in January 1977 by Stanley Watson and John Waterbury aboard *Atlantis II,* these microbes (*Synechococcus*) will show up in nearly all water samples subsequently examined.

In January 1979, the task of writing scientific papers gets easier with acquisition of a Wang word processor, accessible to multiple users simultaneously.

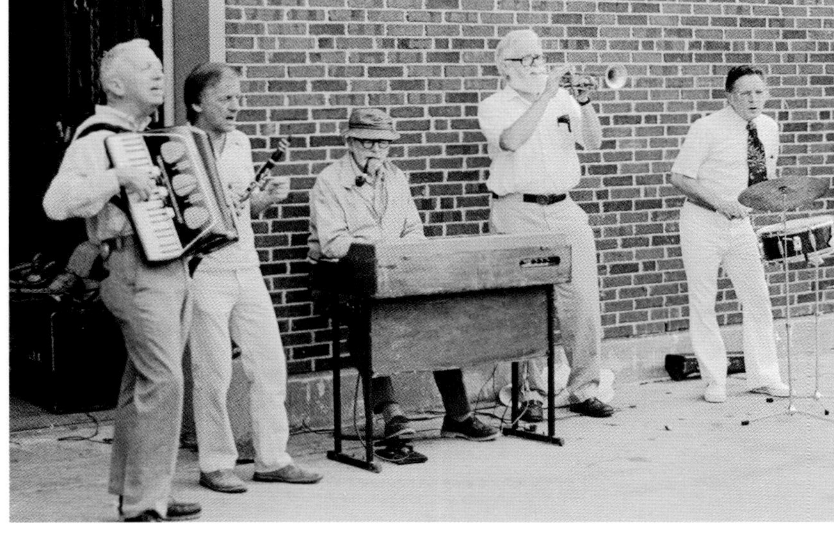

Institution interest in coastal research is reflected during 1979 in planning for a coastal research center and identification of this subject as one of three recommended for special emphasis by an advisory committee made up of Corporation members and individuals from business, government, and academia. The other recommended areas of emphasis are biochemistry and theoretical studies. A center for the analysis of marine systems is also being organized.

Recovery of a POLYMODE array from the Mid-Atlantic Ridge in October 1979 marks a new record for the Buoy Group, a 515-day deployment. The mooring was set during a May 1978 *Atlantis II* cruise. POLYMODE is a collaborative, Russian-American, IDOE program organized to study the role of eddy processes.

In a long-standing tradition, the Woods Hole Marching & Chowder Society helped welcome *Knorr* to homeport in August 1979, following a nineteen-month Pacific voyage. On this occasion, society membership included, from left, Hoyt Watson, Sam Raymond, Asa Wing, Paul Mangelsdorf, and Charlie Innis. *Knorr* had visited 28 foreign ports and noncontiguous territories and played host to 2,211 scientists during Voyage 73.

Earth's Greatest Interface: A Short History of Air-Sea Interaction Research at WHOI

When things went wrong in the early days of developing moored-instrument technology, the only return might be a "wuzzle," like this one being retrieved by Jim Gifford, left, and Bob Heinmiller in the late 1960s. However, the persistence of the WHOI Buoy Group paid off, as successful moorings began to return orderly strings of instruments and eventually approached a 100-percent record of data collection.

The vast interface of ocean and atmosphere constitutes one of the earth's great interactive engines. WHOI scientists have long investigated upper-ocean and atmospheric processes and their influences on one another. Observation of air-sea interaction began with Al Woodcock, a member of the very first *Atlantis* crew, who migrated to science and, during long days at sea, noted similarities in the patterns of seabird flight and the array of seaweed as it aligned with the wind on the surface of the Sargasso Sea.

Woodcock studied low-altitude turbulence and convection and their impact on smoke screens intended to aid troop landings during World War II. Later he was a principal in postwar investigations of air-sea exchanges of heat and moisture in trade-wind areas as well as the study of salt particles as nuclei for raindrops. Woodcock, Andy Bunker, Joanne Malkus (Simpson), and others developed instrumentation for a succession of WHOI-based aircraft devoted to marine meteorology from 1945 to 1970.

Development of moored surface buoys, an extremely difficult task because of the effect of surface motion on mooring components, was pursued with vigor at WHOI during the 1960s and 1970s. Successful mooring efforts gradually overcame the disappointing failures, and the first long-term (two-year) time series of data on atmospheric forcing and upper-ocean response was achieved in the Long-Term Upper Ocean Study east of the Gulf Stream off Cape Hatteras from 1982 to 1984. With further improvements in measuring devices and mooring technology, WHOI researchers took on new challenges in the tropical Pacific and the Arabian Sea while also developing more capable atmospheric, oceanic, and coupled numerical models.

Other tools used for Institution air-sea research include towers built along the shore, on islands, aboard ships, and in shallow waters, including one built in 2000 at WHOI's Martha's

The 1982–84 Long-Term Upper Ocean Study (LOTUS) brought home two years' worth of atmospheric and upper-ocean measurements, the first long-term time series for this type of data. In the photo, the surface mooring has been launched, and the first of a string of current meters to be deployed beneath it is about to go in the water.

Vineyard Coastal Observatory. In another approach, small catamarans bristling with instruments are sent out from ships in the open ocean to avoid ship-based contamination of measurements.

Today WHOI researchers are world leaders in air-sea observations, surface mooring technology, upper-ocean dynamics, and the ocean's role in climate, and they look forward to continuing their pioneering work in air-sea interaction in future decades.

This feature is based on an unpublished 2004 manuscript entitled "WHOI Air-Sea Interaction" by Robert A. Weller, WHOI senior scientist and director of the NOAA/WHOI Cooperative Institute for Climate and Ocean Research.

More recent buoys are equipped with sensors to measure wind speed and direction, air temperature and humidity, incoming solar and infrared radiation, barometric pressure, sea-surface temperature and salinity, and precipitation once a minute. Hourly averaged data is transmitted via satellite and made available to forecast centers and researchers. The photo shows a surface buoy on a mooring deployed in 2003 off northern Chile for a multi-year study of air-sea interaction and upper-ocean variability in an area of persistent low cloud cover. The site is one of a handful of NOAA-funded ocean reference stations.

Fragile—but Robust—Denizens of the Deep:
A Short History of WHOI Research on "Jellies"

Because their bodies have no hard parts, gelatinous animals, commonly called "jellies," have always been fascinating and elusive to naturalists. They were already a favorite subject for Henry Bigelow when he was the curator of the medusae on Alexander Agassiz's 1901 Maldive Island voyage. His drawings of *Porpita lutkeana,* at the bottom of this page, and *Timoides agassizii,* on page 6, are still accurate and beautiful today. During WHOI's early years, Bigelow and his associate Mary Sears continued their research on jellyfish and their relatives.

Decades later, gelatinous animals returned to the list of WHOI research interests when Richard Harbison and Larry Madin joined the Biology Department in the 1970s. Harbison was a biochemist by training but soon became interested in the biology of salps, a type of swimming sea squirt. Madin was finishing his Ph.D. at the University of California, Davis, under William Hamner, who had pioneered new in situ techniques for studying fragile, gelatinous, planktonlike salps in their natural environment. Joining WHOI as a postdoctoral scholar, Madin teamed up with Harbison to introduce "blue-water diving" to WHOI vessels. This approach—scuba diving in open-ocean water to observe and collect delicate jellies that seldom survive net collection—opened up many new lines of research on life history, behavior, energetics, and distribution.

Within a few years the depth limits of scuba diving were overcome by using *Alvin* to extend observation and sampling into much deeper water. Submersible dives to "midwater" depths (near 1,000 meters, or 3,300 feet) in the late 1970s revealed abundant and diverse jellies, including many unknown species that had never been collected with conventional nets. Subsequent midwater dives by Madin, Harbison, and colleagues elsewhere have yielded a hundred or more species new to science but common in the mid-depths of the ocean.

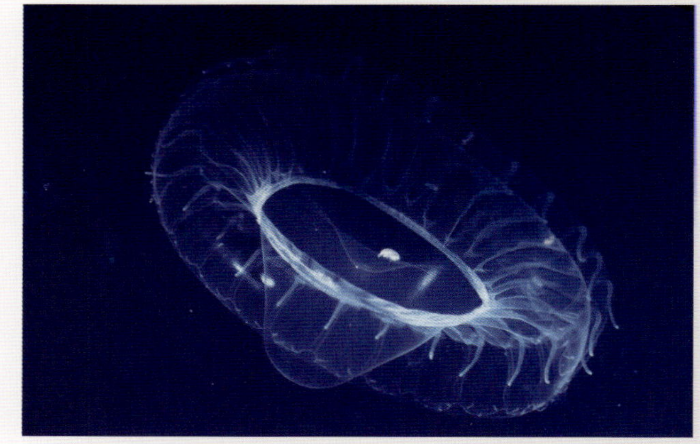

Medusae in the genus *Cunina* live in the open ocean from the surface down to midwater depths of 3,300 feet (1,000 meters). Like most gelatinous animals, this transparent jelly is often inhabited by small crustaceans that use it as a resting place and as a surface to support their juveniles (note the small hitchhiker in the center of the jelly).

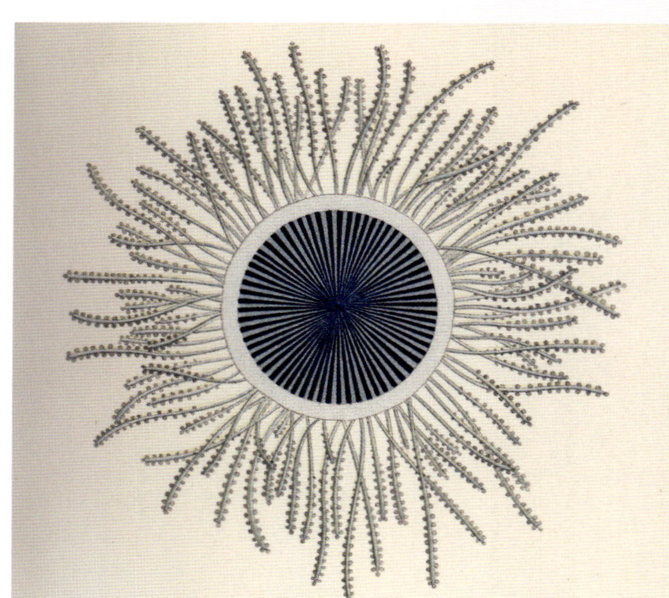

Henry Bigelow drew this medusa, *Porpita lutkeana,* during his 1901 trip to the Maldive Islands.

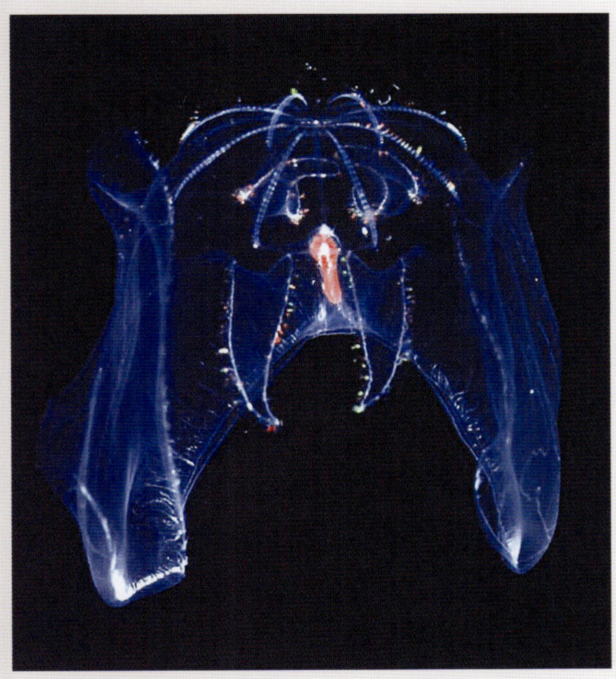

This work has opened a new window on life in the waters of the open sea, revealing and documenting the role of gelatinous animals as integral parts of plankton communities, instead of mere curiosities or nuisances. Henry Bigelow would no doubt be pleased that today's WHOI biologists continue to expand our knowledge of gelatinous animals, documenting their history, role, and effects on large marine systems and processes.

This feature was contributed by WHOI senior scientist Laurence P. Madin, who is also director of the Ocean Life Institute, and science writer Katherine Madin.

As scientists reveal more about the lives of jellies, these creatures become ever more fascinating in their beautiful diversity of form and function. This deep-sea comb jelly, *Bathocyroe fosteri,* is named for *Alvin* pilot Dudley Foster, who collected the first specimens. It is about two inches tall and lives in all oceans at depths below 1,000 feet (300 meters).

Open-ocean scuba diving from research vessels began to reveal more about the world of jellies during cruises in the 1970s. To make their collections and observations and avoid becoming lost in the vast ocean, the divers use a tether system, with one person serving as "safety" at its apex. Underwater lights enable them to make night dives to view and collect jellies when many species are more active and appear in greater numbers.

ATLANTIS
WOODS HOLE

A ship named *Atlantis* is a continuing WHOI tradition. The third *Atlantis,* delivered in 1997, was designed to host deep-sea exploration vehicles, including the submersible *Alvin.*

1980–2005
Springboard for the Future

1980 Construction of McLean Laboratory for geosciences is followed by addition of Coastal Research Laboratory (1981) and Fye Laboratory for chemistry and geochemistry (1983)

1984 A new role for *Atlantis II* as *Alvin* tender lasts until the ship leaves the fleet in 1996

1985 Discovery of the sunken *Titanic* during a *Knorr* expedition shines an international spotlight on Woods Hole

1986 Exploration of *Titanic* by remotely operated vehicle prototype *Jason Jr.* signals a new age of deep-sea exploration

1989 Craig Dorman is named director; World Ocean Circulation Experiment Hydrographic Programme Office is established at WHOI

1993 Robert Gagosian becomes WHOI's seventh director

1995 The first autonomous vehicle debuts, working from *Atlantis II* to map the seafloor on the Juan de Fuca ridge

1997 A new *Atlantis* joins the fleet

2000 Four new ocean institutes enhance cross-disciplinary research and communication of research results; Martha's Vineyard Coastal Observatory becomes operational

2004 The new coastal vessel *Tioga* provides rapid access to coastal research sites; two new laboratory buildings are under construction on the Quissett Campus

Introduction   In its fiftieth year, the Woods Hole Oceanographic Institution is a strong, well-structured organization. It employs 729 people, including an accomplished, productive scientific and technical staff of 208. There are five science departments and a marine policy program. Eighty-two students are enrolled in the MIT/WHOI Joint Program. Three large research vessels (*Atlantis II, Knorr,* and *Oceanus*) and a hardy submersible ply the world ocean, and a coastal vessel works closer to home along with smaller boats. The Quissett Campus accommodates building plans as far as anyone can see into the future. The stage is set for the Institution to continue its leadership role in ocean science for the next quarter century and beyond.

Organization   Over the next twelve years, from 1980 to 1992, the names of two departments change to better reflect their missions: Ocean Engineering becomes the Applied Ocean Physics and Engineering Department in 1989, and Chemistry changes to the Marine Chemistry and Geochemistry Department in 1992. The names of the Biology, Geology and Geophysics, and Physical Oceanography departments remain the same. The Marine Policy & Ocean Management program becomes the Marine Policy Center in 1991. By 2004, the scientific and technical staff numbers 356, there are 142 students in the Joint Program, and WHOI has a total of 957 employees.

WHOI scientists participate in various large oceanographic programs, including the World Ocean Circulation Experiment (WOCE), the Joint Global Ocean Flux Study, the Global Ocean Ecosystems Dynamics Program, Ridge

Knorr encounters an iceberg during the ship's long Transient Tracers in the Ocean (TTO) cruise. TTO is one of many "big science" programs that WHOI scientists and ships have served during the Institution's third quarter-century. Working a zigzag track across the North Atlantic, TTO scientists occupied *Knorr* for seven months in 1981. Fourteen years later, scientists aboard *Knorr* devoted more than a year to the Indian Ocean component of the World Ocean Circulation Experiment (WOCE).

Inter-Disciplinary Global Experiments, the Ocean Drilling Program, the Tropical Ocean-Global Atmosphere program, and others. Offices for administration and operation of several of these programs are based at WHOI. For example, from 1989 to 1996, the WOCE Hydrographic Programme Office was located at the Institution. This office managed the processing, analysis, and quality control of hydrographic measurements collected in what was called the largest physical oceanography program ever attempted.

John Steele serves as director for just over eleven years. He is succeeded in January 1989 by Craig Dorman, who serves until July 1993, when Robert B. Gagosian takes the helm. Gagosian is a twenty-one-year WHOI veteran whose Institution career began with an appointment as assistant scientist in 1972. Following promotion to associate and senior scientist positions, he was named chair of the Chemistry Department in 1982 and moved on to associate director for research in 1987. After six months as acting director, he is elected seventh director in January 1994 and becomes president and director in 2002. Gagosian did undergraduate work at MIT and holds a Ph.D. in organic chemistry from Columbia University. He did postdoctoral work at the University of California, Berkeley, before joining the Institution staff.

John H. Steele, Fifth Director

Director John Steele, center, and Associate Director for Research Derek Spencer, right, welcome Massachusetts governor Michael Dukakis for a visit to WHOI in 1983.

Board Chair Charles Francis Adams, left, and John Steele donned academic regalia for the 1980 commencement held on the WHOI pier.

After twenty-six years on the research staff of the Marine Laboratory in Aberdeen, John Steele left his native Scotland to become fifth director of the Woods Hole Oceanographic Institution in October 1977. An active participant in a number of international oceanographic programs and experiments, Steele brought these affiliations, four years' experience as deputy director at Aberdeen, and excellent scientific credentials, along with his community-oriented wife, Evelyn, to his new post.

Steele made his first visit to Woods Hole in 1958 when Buck Ketchum offered him a six-month fellowship to work on phytoplankton dynamics. Other visits over the next twenty years included service on two Biology Department visiting committees. During the 1970s he was a member of the NSF review panel for IDOE and the steering committee for the Controlled Ecosystem Pollution Experiment, an IDOE program.

His years as director were a time of intellectual and scientific growth for the Institution. Steele and associate director for research Derek Spencer carefully cultivated the Institution's high scientific standards. The first and several subsequent senior scientist chairs were established. Steele, Spencer, and various members of the scientific staff contributed significantly to national and international ocean science in the planning and early organization of global climate change and other large-program studies, including the World Ocean Circulation Experiment, the Joint Global Ocean Flux Study, the Global Ocean Ecosystems Dynamics program, and Ridge Interdisciplinary Global Experiments.

Three new laboratories were constructed with private funding during Steele's directorship: the McLean Laboratory for geosciences, the Fye Laboratory for chemistry and geochemistry, and the Coastal Research Laboratory. An addition to the Clark Laboratory was also initiated. *Atlantis II* was refitted to support *Alvin* operations, there was early planning for a new research vessel, and significant progress was made in the development of remotely operated vehicles.

After stepping down as director in January 1989, Steele continued to serve until 1991 as president of the Corporation, a position he assumed in 1985. He has also contributed to national and international ocean science program planning, and, as senior scientist and scientist emeritus in the Marine Policy Center, he has been involved in a variety of research and policy issues ranging from the conservation of biological diversity to the structure and dynamics of food webs and the analysis of ecological time series.

Ships   Just as the need for more capable, longer-range ships drove the late 1950s transition from the sailing vessel *Atlantis* to the sturdier ships *Crawford* and *Chain,* scientists' demands for longer-range vessels bring significant fleet changes in the late twentieth century.

After nearly two decades as *Alvin*'s tender, the 105-foot (32-meter) *Lulu* proves too small and slow for reaching many desired destinations. WHOI and the funding agencies cast about for a suitable alternative and finally settle on converting *Atlantis II,* already engaged in a midlife refit. Outfitted with a large, hydraulic stern A-frame and a submersible hangar, *Atlantis II* becomes the tender for *Alvin* in January 1984, and a new era in submersible operations begins. *Atlantis II* can launch the submersible in higher sea states than *Lulu* could and has twice the speed, four times the range, three times the science complement, comfortable accommodations, and better labs. The dynamic duo of *Atlantis II* and *Alvin* count many successes over the next thirteen years.

A new era of ocean exploration began in the early 1980s with installation aboard *Atlantis II* of a hangar for *Alvin* and a 41-foot (12.4-meter) hydraulic stern A-frame for handling the sub. For the first time, *Alvin* could be serviced under cover, and submersible operations gained greater geographical range and higher sea-state capability. The photo was taken during sea trials following refit of the ship.

However, by 1996, it is time for *Atlantis II* to retire. With a record of thirty-four years of service, more than a million miles for science, and over eight thousand days at sea, the ship goes out of oceanographic service on July 3, 1996. A new *Atlantis,* constructed by the U.S. Navy (and not numbered because the names of Navy ships bear no numbers), takes its place the following year. At its 1994 launch, this ship was not intended to host the submersible; but again, after considering other options, including transferring *Alvin* operations to *Knorr,* *Atlantis* gets the nod. By 2004, *Atlantis* has proved a highly capable research vessel and submersible tender, working in study areas ranging from the Gulf of Alaska and waters around Easter Island to the Mid-Atlantic Ridge.

One of R/V *Knorr*'s grandest adventures, at least in terms of public attention, occurs in the ship's fifteenth year when its scientific party, led by senior scientist Bob Ballard, brings home the first photographs of RMS *Titanic,* taken by the towed camera system called *Argo.* This "unsinkable" ship's very name has evoked images of wealth and mystery since its fatal encounter with an iceberg in 1912, and a horde of journalists descends on Woods Hole when *Knorr* returns from the *Titanic* discovery cruise in 1985 (and again the following year when *Atlantis II* and *Alvin* visit the *Titanic* site). However, within the oceano-

A *Knorr* refit from 1989–1991 included lengthening the vessel by 33 feet (10 meters) to 279 feet (85 meters). The ship was cut in half at McDermott Shipyard in Amelia, Louisiana, to accommodate installation of a new propulsion system and new construction that would give it greater range, allow a larger science party, and provide more deck and lab space.

Oceanus approaches Woods Hole following a 2001 cruise, as *Atlantis,* in the distance, departs homeport. A 1993–1994 midlife refit for *Oceanus* included expansion of laboratory and science berthing space and construction of a new pilothouse. The original twin stacks were replaced with a single stack.

graphic community, *Knorr* is best known as a workhorse of a ship. To accommodate longer cruises for large global programs, *Knorr* and its sister ship *Melville,* operated by Scripps, are refitted and lengthened in 1989. They indeed well serve programs such as the World Ocean Circulation Experiment and the Joint Global Ocean Flux Study, traveling the world's oceans through the 1990s and into the new millennium.

In 2005, except for *Atlantis,* WHOI's fleet is showing its age. By seagoing standards for research vessels, *Oceanus* and *Knorr* should retire in 2009 and 2015, respectively, and other ships in the U.S. academic fleet are in similar circumstances. Thus considerable planning and discussion take place nationwide regarding the future of ocean science at sea, including the "Access to the Sea Task Force Report" released by WHOI in July 2004. This comprehensive document, compiled by twenty-five members of the Institution's scientific, technical, and operations staff, provides specific recommendations for keeping the Woods Hole Oceanographic Institution at the forefront in four areas: ships and ship operations, vehicles and the National Deep Submergence Facility,* ocean observatories and observing systems, and data management and visualization.

---

* In 2005, the National Deep Submergence Facility comprises *Alvin, Jason II,* the towed camera and sonar sled *Argo II,* and the towed sonar sled *DSL-120.*

New Tools   WHOI's tradition of excellence in ocean instrumentation endures. Scientists and engineers collaborate to improve existing techniques and develop new ones for studying the seafloor, water column, and marine atmosphere as well as their interfaces. For example, recent refinement of mooring operations aims to lower costs, simplify deployment, and extend the lives of moorings that provide meteorological and water column data. One advance features release of data-transmitting packages from a subsurface mooring over a five-year period; another sends a vertical profiler up and down a mooring wire to make continuous measurements of temperature, salinity, and water velocity for months at a time. The technology for moored sediment traps that began to bring new knowledge in the late 1970s of particulate matter descending through the ocean has migrated to neutrally buoyant, floating sediment traps. Other stationary instruments include tripods that suspend current meters for study of small-scale, near-bottom currents or devices to extract sediment or pore water from beneath the seafloor for chemical analysis. Towed biological instruments range from large-net environmental sensing systems for sampling at several levels during a single tow to video, optical, and acoustic devices that sense or capture images of plankton populations. There are advances in tagging, listening devices, and other means of studying marine mammals.

Deep-sea vehicles broaden horizons for ocean scientists beginning in the 1980s. Led by Bob Ballard, scientists and engineers in the Deep Submergence Laboratory (DSL) begin to design a remotely operated vehicle (ROV) to provide a virtual human presence on the seafloor. They call the vehicle *Jason*, for the mythical Greek adventurer and ocean explorer, and design it for operation from a specially outfitted, portable van aboard ship.

The DSL group builds a 28-inch (70-centimeter) ROV prototype called *Jason Jr.* that *Alvin* carries to the *Titanic* site. The tiny ROV explores the sunken ship's interior on a 200-foot (60-meter) tether that transmits commands from an ROV pilot inside the sub. The next step is to build a full-scale *Jason* in 1987. The Navy-funded vehicle makes its debut at sea in 1988, and over the next four years its developers struggle with the inevitable engineering challenges a new system presents, as well as scientists' reluctance to invest their hard-earned funding in an unknown tool. One success is the 1989 initiation of The JASON Project, Bob Ballard's new venture

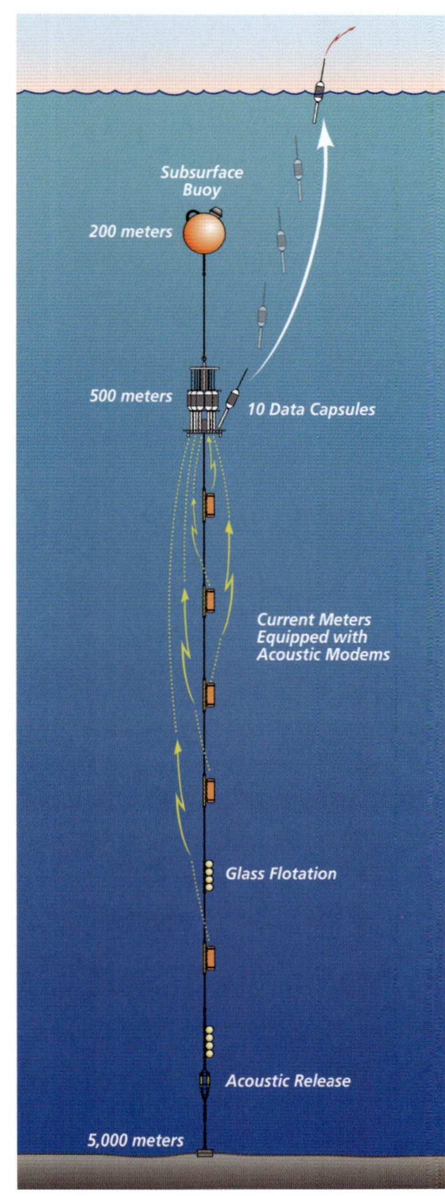

"Ultramoor" is an early twenty-first century advance in technology whose objective is to extend mooring life from two to about five years. Current meters and other instruments placed along the mooring line are equipped with acoustic transmitters that transfer the data they collect up the line for loading onto data capsules. One capsule is released about every six months to ascend to the surface and transmit the stored data via satellite to researchers ashore.

When WHOI's Deep Submergence Laboratory (DSL) was established in 1983, Bob Ballard, right, Dana Yoerger, left, and Ken Stewart wrote in the annual report that technological advances in video imagery, robotics, fiber optics, and microprocessing held great promise for advanced undersea-exploration vehicles. The small vehicle pictured, called *RPV-430,* was built by Benthos, Inc., for use as a DSL test bed for research on supervisory control, human-machine interface design, and robotics as the group developed the 20,000-foot (6,000-meter) remotely operated vehicle *Jason.* Now, twenty-two years later, the Institution operates a fleet of remotely operated and autonomous vehicles that are based on this work.

in communicating science to the public. It features live broadcasts from the seafloor and ship's deck to schoolchildren gathered at museums and other sites around the country, including WHOI. (The JASON Project soon becomes a separate nonprofit foundation that contracts with the Institution for DSL services.)

By the mid-1990s, scientists are reporting successful *Jason* expeditions to their colleagues, and the ROV becomes an important tool for deep-ocean science. In 2002, the second-generation *Jason II* brings U.S. ocean scientists a more capable vehicle, and the DSL team also builds a sister vehicle, *Isis,* for Great Britain.

With its exploration of RMS *Titanic* in 1986, WHOI's prototype remotely operated vehicle *Jason Jr.,* at right, proved that Deep Submergence Laboratory scientists and engineers were on the right track.

Jason II, shown in the inset picture, is similar in appearance to but larger than the first *Jason.* Based on ten years of experience with the original vehicle, WHOI's Deep Submergence Laboratory team designed *Jason II* to dive deeper (to 21,450 feet or 6,500 meters) and offer greater maneuverability, dexterity, payload, and power.

Surveying Seafloor Tragedies

Though never a major focus of WHOI work, searching the seafloor for wrecks has been a recurring theme, often with the purpose of proving deep-sea technology. The earliest wreck searches were conducted during World War II, using echosounders and underwater cameras to help the Navy locate sunken ships and downed planes. In 1962, an *Atlantis II* biology cruise was diverted to the frustrating search for the nuclear submarine *Thresher* (see page 117). The next big event in this line was exploration of RMS *Titanic* during *Knorr* and *Atlantis II–Alvin* cruises in 1985 and 1986. Eleven years later the Deep Submergence Laboratory team conducted what must be the most thorough investigation ever undertaken of a sunken ship using two towed vehicles and the remotely operated vehicle *Jason*.

Loss of large bulk carriers had been a continuing concern in the maritime industry, and the British government wanted to know what caused the 1980 loss in the Pacific Ocean of the 964-foot (292-meter) bulk ore carrier *Derbyshire* and the forty-four people aboard. Impressed by WHOI's long history of deep-sea exploration, success in search and survey missions, and development of high-quality underwater photography, U.K. officials engaged the Institution for a survey. Analysis of 137,000 digital color and high-definition video images collected by the WHOI vehicles over two months in 1997 showed that the tragedy was not caused, as feared, by massive structural failure but rather by flooding of forward spaces through an open hatch and damaged ventilators that initiated a series of flooding and collapse events. A British official summed up the impact of this survey by saying, "No longer need ships vanish without trace or explanation." [147]

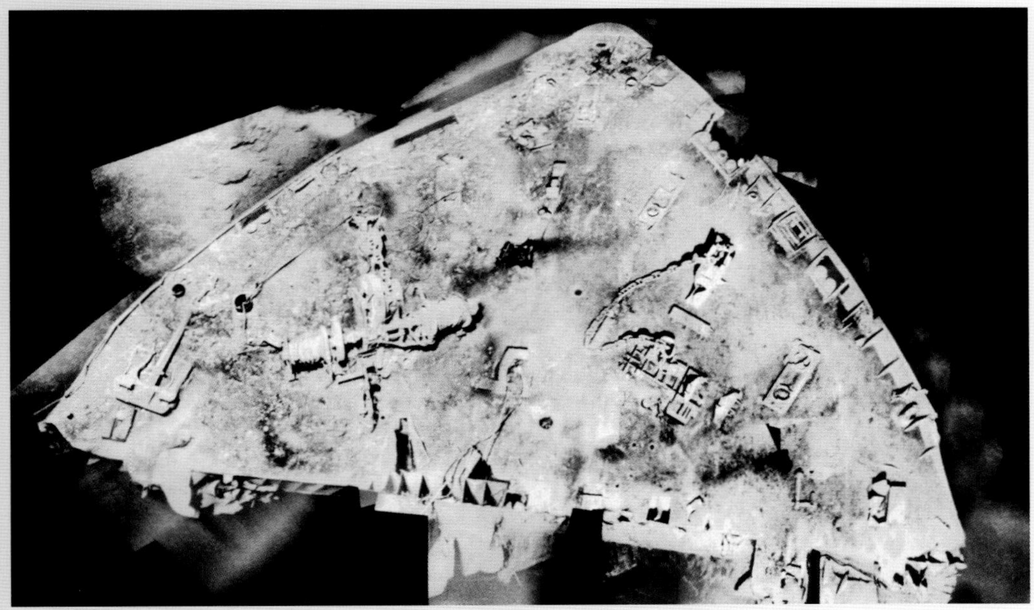

This mosaic of the bulk carrier *Derbyshire*'s bow section, comprising fifty electronic images from the towed camera and sonar sled *Argo,* shows the only part of the ship found essentially intact, which indicated that the bow was flooded before sinking. The rest of the vessel was fragmented into more than two-thousand pieces.

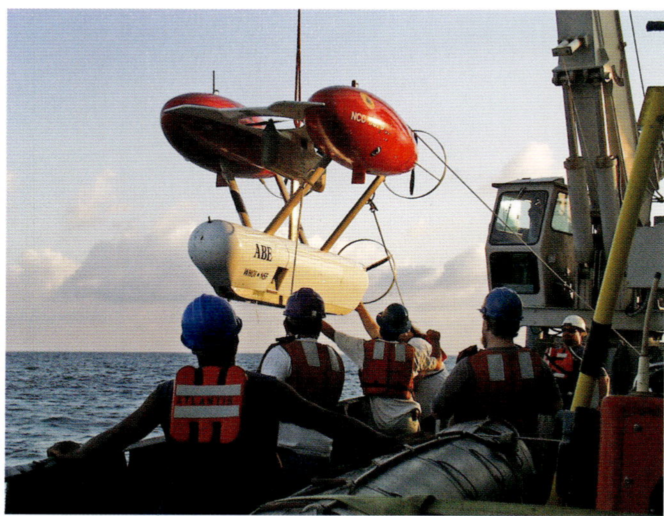

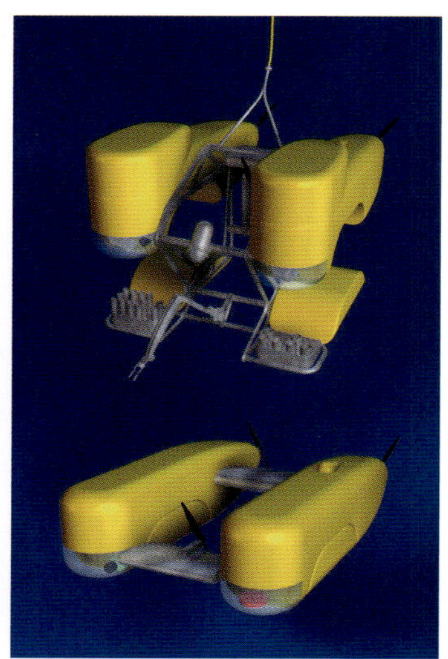

As the tethered vehicle gains acceptance, a new untethered vehicle begins to take shape. The design of the *Autonomous Benthic Explorer (ABE),* the first deep-sea autonomous underwater vehicle (AUV) in the U.S. ocean science community, employs many lessons learned with the ROV. However, its design builds the controls into the vehicle so that it can be programmed to operate independently near the seafloor. *ABE* makes its debut in the summer of 1995. While *Alvin* explores the area during the day, *ABE* makes night dives to map seafloor magnetic anomalies at a depth of 8,000 feet (2,400 meters) on the Juan de Fuca Ridge off the coast of Oregon. Over the next several years, demands for *ABE*'s services increase. In 2005, a new autonomous vehicle known as *Sentry* is under construction, and several other autonomous vehicles have taken their places in the constantly expanding suite of WHOI's ocean exploration tools. They include deep and shallow vehicles of *REMUS* (*r*emote *e*nvironmental *m*onitoring *u*nit*s*) design, with assignments ranging from the original ocean science purposes

to mine detection. There is an autonomous surveying and imaging vehicle called *SeaBED,* and a special under-ice autonomous vehicle is being developed.

In another category of vehicles, gliders are proving invaluable. They travel a preprogrammed, sawtoothed course from the surface to 3,300 feet (1,000 meters) for a month or more while reporting a variety of data to laboratories ashore via satellite. These vehicles hold great promise for collecting data from around the globe, including remote and hostile environments.

Development of AUVs is part of today's response to the need scientists have always felt for continuous monitoring of ocean phenomena. In his 1990 comments as associate director for research, Bob Gagosian writes, "Collecting data 24 hours a day, 365 days a year has become a very important objective since order-of-magnitude changes can occur in a very short time frame and completely change our conclusions concerning the dominant process controlling the phenomenon being studied." The concept of ocean observatories evolves to fill this need, with the expectation that AUVs will play an important role in detecting significant phenomena and collecting data over long periods of time. WHOI staff contribute to an early observatory effort called LEO-15 (Longterm Ecosystem Observatory—15 meters) established in 1996 by the Rutgers University Institute of Marine and Coastal Sciences off the coast of New Jersey. In 2000, assembly of the initial components of WHOI's own Martha's Vineyard Coastal Observatory concludes with installation of a meteorological mast. Today, Institution scientists and engineers are part of visionary teams planning a number of other global ocean-observing systems that promise major advances in understanding planet Earth.

Alvin made its 4,000th dive in 2004, its 40th year. One of only five deep-sea research submersibles in the world, *Alvin* is considered the most active, recently averaging 175 dives per year. While the sub has been carefully maintained and improved over the years (only the name is original), its current depth rating of 14,764 feet (4,500 meters) leaves more than 30

Located on the south shore of the island, facing the open ocean, the Martha's Vineyard Coastal Observatory provides ideal conditions for studying coastal processes. A wide variety of sensors attached to a near-shore seafloor node, a meteorological mast, and this tower located 1.8 miles (3 kilometers) offshore send data by cable to a small beachfront laboratory, where they can be distributed via the Internet.

The second *Asterias,* in service for science since 1979, was replaced in 2004 by *Tioga,* a larger, faster, more capable coastal vessel. *Tioga* was designed to provide ready access to a broader range of study areas, including the Martha's Vineyard Coastal Observatory.

Dedicated in 1980 to long-time board chair Noel McLean, this Quissett Campus building was designed by New York architects Cain, Farrell & Bell to house geoscience laboratories along with a fifty-year collection of sediment cores, rocks, and other seafloor samples. The Institution's data library and archives are also located here. The Data & Earth Sample Center building, where some of these facilities had been located since it was built in 1971, was then leased to the Atlantic–Gulf of Mexico Branch of the U.S. Geological Survey.

percent of the seafloor outside its range. In August 2004, NSF announces funding for a replacement for *Alvin;* its characteristics will include better fields of view and ergonomics for the pilot and diving scientists and a number of other improved systems, but, most important, a depth capability of 21,450 feet (6,500 meters). This will offer human access to 99 percent of the seafloor.

WHOI Mission Statement

Adopted in 1990

The Woods Hole Oceanographic Institution is a private, independent, not-for-profit corporation dedicated to research and higher education at the frontiers of ocean science.

Its primary mission is to develop and effectively communicate a fundamental understanding of the processes and characteristics governing how the oceans function and how they interact with the Earth as a whole.

To fulfill this mission, WHOI must successfully:

- *Recruit, retain, and support the highest quality staff and students and provide an organization which nurtures creativity and innovation.*

- *Stress a flexible, multidisciplinary, and collaborative approach to the research and education activities of its staff within an equitable working environment.*

- *Promote the development and use of advanced instrumentation and systems (including ships, vehicles and platforms) to make the required observations at sea and in the laboratory.*

- *Make the results of its research known to the public and policymakers and foster its applications to new technology and products in ways consistent with the wise use of the oceans.*

- *Secure the essential resources to sustain these activities, a responsibility which the Trustees and Corporation Members must jointly share with management and staff.*

It is the goal of the Institution to be a world leader in advancing and communicating a basic understanding of the oceans and their decisive role in addressing global questions.

Craig Dorman, Sixth Director

Craig E. Dorman, a 1972 graduate of the MIT/WHOI Joint Program, returned to his alma mater as director in January 1989. He had served in the U.S. Navy for twenty-six years following graduation from Dartmouth College with a degree in geography. He had earned his master's degree in oceanography from the Naval Postgraduate School and Ph.D. in physical oceanography from the MIT/WHOI Joint Program during Navy service. His naval career included operational experience with underwater demolition teams and the SEALS, followed by more than ten years' involvement with the research and development aspects of antisubmarine warfare. His last Navy position had been rear admiral and program director of antisubmarine warfare in the Space and Naval Warfare Systems Command.

During his first year as director, Craig Dorman (right) chats with Barrett (Buzzy) McLaughlin aboard *Atlantis II*, one of several WHOI ships McLaughlin served as chief engineer.

During his four years as WHOI director, Dorman worked to develop international alliances, especially with Russia, and also to improve industrial liaison, while strengthening Institution relationships with federal agencies. This was a time of aggressive federal auditing of nonprofit institutions, and his administration weathered this along with a problematic refit for R/V *Knorr* and difficulty with implementing a computerized financial system. At the same time, WHOI's federal research funding increased by 26 percent, and board chair Guy Nichols said, "The Institution is in an excellent position to continue as the nation's premier oceanographic institution." [148]

Dorman placed great value on creating a supportive environment for WHOI scientists and students. Between the deanships of Charles Hollister and John Farrington, he served as dean for a few months. He often moved about the Institution, encouraging and engaging employees at all levels of the workplace. He and his wife, Cynthia, enjoyed hosting social gatherings, including popular monthly "soup and bread suppers" for employees.

In his departing *Director.gram* (#44), Dorman noted that a 1992 visiting committee had found "our research and educational programs extremely strong" and that he felt he was leaving a place that "knows where it's going, is on track, in control of its future, on the rise, and ready to adjust to whatever comes next."

Craig Dorman presided at festivities celebrating the submersible *Alvin*'s twenty-fifth anniversary in 1989.

After his tenure at WHOI, Dorman held U.S. Navy positions for several years, and in 2005 he was vice president for research at the University of Alaska.

Woods Hole village is the cluster of buildings at upper left in this September 2004 aerial view. Extensive construction is apparent on WHOI's Quissett Campus in the foreground. The Environmental Systems Laboratory is at the bottom of the photo below the leaching field for a new wastewater treatment plant; the McKee Ballfield will be reinstated in this area in 2005. Other Quissett Campus facilities, beginning with the Clark South/Clark complex above the leaching field and moving counterclockwise, are the renovated Central Plant and attached new Marine Research Facility, the McLean Laboratory, steel framing for the new Biogeochemistry Building with the Carriage House conference room to its left, Fenno House (with Bell House among the trees to its left), the Fye Laboratory, the Shepley tennis and volleyball courts, and the Rinehart Coastal Research Laboratory.

The National Ocean Sciences Mass Spectrometry Facility was installed during 1989 in an addition to the McLean Laboratory. At its dedication in 1990, the NSF-funded facility was one of only seven of its kind in the world, the most sophisticated and technologically advanced of these, and the only one designed specifically for ocean science research. Since then, it has provided the ocean sciences community with carbon-14 dating, offering quick response time and requiring only small samples of seawater, sediment, and other substances important to paleoceanography and other fields. In the photo, Bob Schneider loads a carousel of samples for analysis.

Facilities   Except for construction about 1990 of an addition to the Clark Laboratory, known as Clark South, and of a dozen housing units on the Quissett Campus, few changes to the Institution's physical plant occur near the end of the twentieth century. As the new millennium begins, a need for laboratory space inspires development of a master plan for plant improvement. Implementation of this plan begins with construction in 2003 of a ring road around most of the Quissett Campus science buildings to create a central laboratory-and-pedestrian area. By the middle of 2004, two new buildings are going up within the ring, one to bring scientists from various disciplines together in a biogeochemistry group and another that will offer laboratories for the marine mammal and other research teams. The two buildings will provide more than 70,000 square feet (6,300 square meters) of new lab space, a 30 percent increase. Additions to the McLean Laboratory to expand seafloor sample storage and the National Ocean Studies Accelerator Mass Spectrometry Facility

(NOSAMS) are also initiated, along with infrastructure improvements. In the village of Woods Hole, the first stages of improvement include enlarging the Blake Building for expansion of the Deep Submergence Laboratory.

In addition to the National Deep Submergence Facility and NOSAMS, WHOI is the site of the Northeast National Ion Microprobe Facility, which measures trace elements and stable radiogenic isotopes in rock samples. The Geology and Geophysics Department is home to one of three sites for the NSF-sponsored National Ocean Bottom Seismic Instrument Pool; the WHOI, Scripps, and Lamont facilities maintain and operate ocean bottom seismometers and hydrophones for studies of structures and processes beneath the seafloor. The seventy-five instruments in the WHOI pool include the latest of several generations of these tools developed by WHOI scientists beginning in the 1970s.

Women Oceanographers Although the recruiting pool remains limited, important strides are made in recruiting female scientific staff in the 1990s. For example, in 1988, seven of the 100-member scientific staff are women; by 2003, 24 of the 135-member scientific staff are female, and the pattern is similar for the technical staff. The last two departments name their first female scientific staff members in 1986 when Cheryl Ann Butman and Kathryn Kelly are named assistant scientists in ocean engineering and physical oceanography, respectively. Jean Whelan is the first woman to achieve the highest rank on the technical staff when she is named

senior research specialist in 1983. Susan Humphris becomes the first permanently appointed female department chair in 2004, heading the Geology and Geophysics Department.

There is a progressive increase in female applicants to WHOI academic programs beginning in the 1970s. It is first noticeable in the undergraduate Summer Student Fellow Program, followed by the MIT/WHOI Joint Program, and more recently, at the postdoctoral level.

Ocean Institutes   In 2000, formation of four ocean institutes constitutes an important new initiative designed to encourage interdisciplinary scientific programs, to focus WHOI research attention on ocean-related issues that have substantial impact on society, and to communicate the results in ways that make them understandable and available to the public and policy makers. The Coastal Ocean Institute joins the Rinehart Coastal Research Center to examine issues concerning coastal processes. The Deep Ocean Exploration Institute sponsors studies of processes that shape Earth's surface and regulate ocean chemistry. The Ocean and Climate Change Institute pursues greater understanding of the ocean's role in climate change, and the Ocean Life Institute fosters research on key questions in ocean biology and ecology. The NOAA/WHOI Cooperative Institute for Climate and Ocean Research, formed in 1998, also enhances interdisciplinary coastal, climate, and ecosystem research.

Funding Ocean Science   There have been many ups and downs in federal funding over the years, including during what oceanographers now consider the "golden years," the quarter of a century following the *Sputnik* launches. In the early days, the Institution was able to operate quite well on Rockefeller Foundation annual support and endowment income, then later with block Navy funding, and later still with grant and contract income from ONR, NSF, and other agencies. Federal agency funding tightens considerably during the 1980s, and by 1989 associate director for research Bob Gagosian observes in the annual report that "Competition for federal funding sources dramatically increased as the decade came to a close. As a result, our Scientific and Technical Staffs have had to sharpen their focus and spend more time trying to market and sell their science." This trend continues through the 1990s, and today ocean scientists spend a great deal of their time writing and rewriting proposals for ever-shrinking federal dollars.

As a result, private funding is of growing importance to the ocean sciences. The Institution undertakes its first capital fund-raising effort in about 1970, when major gifts enhance education and building programs. A sustained effort at raising private funds follows, and a formal capital campaign yields more than $50 million in the 1990s. In 2005, the Institution is about at the midpoint of an ambitious campaign to raise more than $200 million to support research and education under the ocean institutes, new research facilities under construction, and the new coastal vessel *Tioga*.

The Next Seventy-five Years

In its seventy-fifth year, the Woods Hole Oceanographic Institution is a robust private, nonprofit, scientific enterprise. New awards are often announced for its outstanding students and scientists, who fare remarkably well in the increasingly difficult arena of federal funding competition. New facilities being constructed ashore will enhance their work, and plans are well under way to maintain Institution leadership in a variety of marine operations. The next seventy-five years promise new, noteworthy achievements in ocean research, education, and exploration.

Members of the WHOI directorate gathered in the new wing of Fenno House for a portrait in August 2004. President and Director Bob Gagosian is at center, flanked on the right by Bob Detrick, vice president for marine facilities and operations, and on the left by Jim Kent, director of communications. Seated left to right are Jim Luyten, executive vice president and director of research; John Farrington, vice president for academic programs and dean; Carolyn Bunker, vice president for finance and administration; and Dan Stuermer, vice president for board relations and director of development. Director of Government Relations Terry Schaff was not present for the photo. In 1988, the director's office was relocated from the Clark Laboratory to Fenno House, where a new wing was added in 1999. It houses the president and director's office and staff, the communications and development groups, and an employee lunchroom called the Buttery.

Appendix I

Woods Hole Oceanographic Institution Chairs and Presidents

Frank R. Lillie

Henry B. Bigelow

Arnaud C. Marts

Raymond S. Stevens

Frank R. Lillie served as president of the Corporation from the Institution's founding in 1930 until 1940. He had chaired the National Academy of Sciences' committee recommending establishment of an East Coast laboratory that became the Woods Hole Oceanographic Institution. He also served the Marine Biological Laboratory in Woods Hole as director for seventeen years and for another seventeen years as its president. An embryologist and cell biologist, Lillie spent most of his career based at the University of Chicago.

Henry B. Bigelow (WHOI's director from 1930 to 1940) succeeded Lillie as president. He became chair of the Board of Trustees in 1950 when **Arnaud C. Marts** was named president of the Corporation. Marts brought extensive experience in the nonprofit sector as an organizer and president of Marts & Lundy, Inc., financial counselors for philanthropic institutions, and as president from 1935 to 1945 of Bucknell University. In 1955, **Raymond S. Stevens,** vice president and director of Arthur D. Little, Inc., and an active Massachusetts Institute of Technology alumnus, succeeded Marts and served until 1960.

Noel B. McLean

Paul M. Fye

In 1961, **Noel B. McLean** began twelve years of service as chair of the Board, and Director **Paul M. Fye** assumed the title of president of the Corporation. McLean had broad experience in the marine industry with the Marine Division of Bendix Aviation and with Edo Corporation, where he was president and chair of the board. He received the Navy Distinguished Public Service Award in 1972 "for outstanding service . . . in the fields of sonar development, antisubmarine warfare technology, and oceanographic research."

Charles F. Adams

Charles F. Adams was chair of the board from 1973 until 1984. He was continuing a family tradition when he replaced his father as a WHOI Corporation member in 1954. (His father was secretary of the Navy under Calvin Coolidge and a close friend of Henry Bigelow.) Following seven years in the investment business, Adams joined Raytheon as executive vice president and later became president and chair of the board.

John H. Steele

Fye continued as president of the Corporation until 1985 when then-director **John H. Steele** became president.

Guy W. Nichols

Guy W. Nichols was chair of the board from 1985 until 1994. Nichols joined New England Electric following his 1947 graduation from the University of Vermont and eventually served as president, chief executive officer, and chair, building the company into the strongest and best-managed utility in New England. He was succeeded by **Frank V.**

Frank V. Snyder

Snyder, a World War II submariner (whose first contact with WHOI was when an Institution representative installed a bathythermograph on his vessel). His business experience included law practice, the oil industry, and cofounding The Stratton Corporation to develop the Stratton Mountain ski resort.

James M. Clark

James M. Clark followed Steele as president of the Corporation in 1991, and his title was changed to chair of the Corporation in 2001, when Robert Gagosian was named president as well as director. Clark's family had long supported WHOI; the Clark Laboratory was named for his parents upon its completion in 1974. His career in finance included executive positions with Shearson Lehman Brothers, Inc., and Hornblower & Weeks-Hemphill, Noyes, Inc., where he was general partner and CEO; he also has many philanthropic board affiliations.

Thomas B. Wheeler

Thomas B. Wheeler succeeded Clark as Corporation chair in 2004. His career with the Massachusetts Mutual Life Insurance Company spanned more than thirty years, ending as chair and CEO. Previous affiliations included chair of the Massachusetts Business Roundtable, Jobs for Massachusetts, and the Massachusetts State Olympic Committee. He also served as chair of the Board of Trustees at Springfield College.

James E. Moltz

The present chair of the WHOI board is **James E. Moltz,** whose long career in finance has been based at Cyrus J. Lawrence and Sons, for which he served as chair and president until its purchase by Deutsche Bank Securities Corporation, where he was chief investment officer. Moltz, a former director of the New York Stock Exchange and of the Securities Industry Association, is currently vice chairman of ISI.

The Henry Bryant Bigelow Medal
Awarded to those who make significant inquiries
into the phenomena of the sea.

1960   *Henry Bryant Bigelow*
In recognition of his monograph from his collec-
tions on the Albatross Expedition to the Eastern
Tropical Pacific, his comprehensive investigations
of the biology and physical oceanography of New
England waters, and in gratitude for his leadership
in the development of the ocean sciences in the
United States.

1962   *John C. Swallow*
In recognition of his work with the Swallow Float,
which revolutionized our conceptions of the character of the deep water motions.

1964   *Bruce C. Heezen*
In recognition of his physiographic studies of the
deep ocean and of related marine sedimentary
and tectonic processes.

1966   *Columbus O'Donnell Iselin*
In recognition of his contributions to our understanding of the circulation of the North Atlantic
Ocean, for his long and valuable leadership of the
Woods Hole Oceanographic Institution, and for
the profound influence he has had on the development of Oceanography in the United States.

1970   *Frederick John Vine*
In recognition of his imaginative and sound contributions to man's understanding of the formative processes active within the earth.

1974   *Henry Melson Stommel*
In recognition of his contributions to our understanding of ocean circulation and dynamics, for
his continuing intellectual leadership in theoreti-

cal and experimental marine science, and for his
profound influence on the advance of oceanography throughout the world.

1979   *Wolfgang Helmut Berger*
In recognition of his creative contribution to paleoceanography by opening the doors of perception
on the controlling factors governing carbonate
sedimentation in the oceans and for providing us
with a unifying conceptual model for interpreting
the geological evolution of ocean basins.

1980   *Holger Windekilde Jannasch*
In recognition of his creative contributions to
marine microbiology by providing us with an
understanding of the fundamentals of microbial
processes in the sea and the dynamics of oceanic
food chains.

1984   *Arnold Lewis Gordon*
In recognition of his dedication in completing the
Antarctic Circumpolar survey and for his continu-
ing scientific leadership in Antarctic oceanography.

1988   *Hans T. Rossby & Douglas C. Webb*
In recognition of their creative contributions to
ocean technology and oceanography, particularly in
the development of the SOFAR Float and advancing
our knowledge of Lagrangian Ocean Dynamics.

1992   *Alice L. Alldredge & Mary W. Silver*
In recognition of their creative contributions to
biological and chemical oceanography, particularly
in demonstrating the importance of "marine snow"
as a major contributor to the vertical flux of particulate material throughout the world's oceans.

1997   *William J. Jenkins*
In recognition of his creative contributions to the
understanding of ocean circulation, mixing, and pro-

ductivity using transient tracer distributions, to the development of the tritium-helium dating technique, and to the understanding of outgassing from the solid earth, which created the ocean and atmosphere.

2004   *David Michael Karl*

In recognition of his contributions to microbial oceanography, especially the development and leadership of long-term, integrated studies of chemical, physical, and biological variations in oceanic environments.

The Bostwick H. Ketchum Award

Honoring an internationally recognized scientist who demonstrates an innovative approach to coastal research, leadership in the scientific community, and attention to the effects of marine pollution on the coastal environment and society.

1984	*Edward D. Goldberg*
1985	*Thomas H. Pearson & Michael Moore*
1986	*John S. Allen*
1988	*Alasdair McIntyre*
1989	*William Boicourt*
1990	*Daniel Lynch*
1992	*Scott Nixon*
1995	*Christopher Martens*
1996	*Ronald Smith*
1999	*Willard Moore*
2002	*Nancy Rabalais*
2003	*John Farrington*

The Henry Stommel Medal in Oceanography

For fundamental and enduring contributions to observing and understanding ocean processes.

1994   *John Crossley Swallow*

In recognition of his exemplary seagoing oceanographic studies of North Atlantic, Mediterranean, and Indian Ocean circulation.

2000   *Russ E. Davis*

In recognition of his pioneering development of autonomous floats and their use to determine the ocean circulation.

The Mary Sears Woman Pioneer in Oceanography Award

For a woman who has provided significant scientific leadership in understanding our marine environment and has provided inspiration and/or opportunity for other women in marine sciences.

2002   *Mary Wilcox Silver*

In recognition of her contributions to our understanding of the marine ecology of plankton, her inspiration to other marine biologists, and her mentoring skills.

Prior to formal establishment of the Mary Sears award in 1999, the Women's Committee of the Woods Hole Oceanographic Institution had informally honored four Women Pioneers in Oceanography. They are:

1994	*Mary Sears*
1995	*Betty Bunce*
1996	*Ruth Turner*
1999	*Marie Tharp*

Recommended Reading

For those interested in learning more about WHOI history, four books are especially recommended from among the sources listed in the bibliography. Marine historian Susan Schlee's *The Edge of an Unfamiliar World* (1973) offers a broad look at the history of oceanography in the nineteenth and twentieth centuries. Schlee also tells the lively tale of WHOI's first research vessel *Atlantis* in *On Almost Any Wind* (1978), a volume full of early sea stories that capture the work ethic, cooperation, and spirit of adventure that were and are vital to progress in ocean science. Gary Weir, head of the Contemporary History Branch of the Naval Historical Center in Washington, D.C., relates the now-declassified story of the important role played by U.S. oceanographic institutions in World War II and beyond in *An Ocean in Common* (2001), and the first twenty-five years of the research submersible *Alvin* are vibrantly chronicled in science writer Victoria Kaharl's *Water Baby* (1990). One only needs to read her account of three men's miraculous escape when *Alvin* sank in 1968 to know that the Woods Hole Oceanographic Institution is truly blessed.

Endnotes

Quotations not endnoted are drawn from the published WHOI annual report for the year under discussion.

1. Eric L. Mills, e-mail to author, June 23, 2004.
2. Frank R. Lillie, *The Woods Hole Marine Biological Laboratory* (Chicago: University of Chicago Press, 1944), 37.
3. Henry B. Bigelow, *Memories of a Long and Active Life* (Cambridge, Mass.: The Cosmos Press, 1964), 9.
4. Susan Schlee, *The Edge of an Unfamiliar World: A History of Oceanography* (New York: Dutton, 1973), 255.
5. Bigelow, *Memories*, 23.
6. Harold L. Burstyn, "Reviving American Oceanography: Frank Lillie, Wickliffe Rose, and the Founding of the Woods Hole Oceanographic Institution," *Oceanography: The Past*, M. Sears and D. Merriman, eds. (New York: Springer-Verlag, 1980), 58.
7. Schlee, *Edge*, 180.
8. Burstyn, *Oceanography: The Past*, 58.
9. Ibid.
10. Frank R. Lillie to Wickliffe Rose, December 15, 1927, folder 3668, box 358, 909 Oceanography 1927–1954, Rockefeller Archive Center.
11. Lillie, *Marine Biological Laboratory*, 177.
12. Burstyn, *Oceanography: The Past*, 62.
13. Roy Acheson, *Wickliffe Rose of the Rockefeller Foundation: 1862–1914, The Formative Years* (Cambridge, Eng.: Killycarn Press, 1992), 87.
14. From "Record of Interviews with Doctor Rose": Doctor Frank R. Lillie (for luncheon at Hotel Del Prado), Chicago, Illinois–October 5, 1927, folder 3668, box 358, 909 Oceanography 1927–1954, Rockefeller Archive Center.
15. Susan Schlee, "Real Men Won't Winter on Cape Cod: A Bit of Unofficial W.H.O.I. History," *Woods Hole Notes*, February 1973, 2.
16. Frank R. Lillie to Wickliffe Rose, May 11, 1927, folder 3668, box 358, 909 Oceanography 1927–1954, Rockefeller Archive Center.
17. Burstyn, *Oceanography: The Past*, 63–64.
18. Henry Bigelow and William W. Welsh, *Fishes of the Gulf of Maine. Bulletin of the United States Bureau of Fisheries* XL, Part 1 (Washington, D.C.: Government Printing Office, 1925), between pp. 506–507.
19. Michael Graham. "Henry Bryant Bigelow," *Deep-Sea Research* 15 (1968): 127.
20. Vaughan to Bigelow, "H.B. Bigelow correspondence, Special Subjects 1913–1930," HUG 4212.10, Harvard University Archives.
21. Henry B. Bigelow, "A Developing View-Point in Oceanography," *Science* 71, no. 1830 (1930): 86.
22. Bigelow to Sverdrup, January 22, 1930, "H. U. Sverdrup," HUG 4121.7, Box 10, Henry Bryant Bigelow General Correspondence 1930s and 1940s, Harvard University Archives.
23. Bigelow to Redfield, November 22, 1930; February 7, 1931; and March 4, 1931. Redfield to Bigelow, cable, no date. AC-9, "Redfield, A.C. 1928–1935," Office of the Director, WHOI Archives.
24. Alfred C. Redfield, "Henry Bryant Bigelow: October 3, 1879–December 11, 1967," *Biographical Memoirs* 48 (Washington, D.C.: National Academy Press, 1976): 50–81.
25. Iselin to Bigelow, February 8, 1930, AC-9, "Iselin, Columbus (2 of 2), 1930–1935," Office of the Director, WHOI Archives.
26. *The Woods Hole Oceanographic Institution Report for the Years 1930–32*, 15.
27. Bigelow to Iselin, February 18, 1930, and Iselin to Bigelow, February 20, 1930, AC-9, "Iselin, Columbus (2 of 2) 1930–1935," Office of the Director, WHOI Archives.
28. Columbus O'Donnell Iselin, "Personal History" (unpublished, about 1965), pp. 36–39, MC-16, Box 32, Office of the Director, WHOI Archives.
29. Graham, "Henry Bryant Bigelow," 125.
30. Bigelow, "A Developing View-Point in Oceanography," 86.
31. Mary Sears, interview by Victoria Kaharl, September 29, 1989, transcript p. 6, WHOI Archives.
32. Bigelow statement regarding his view of the role of an institution director, Papers of Henry Bryant Bigelow, General correspondence, 1950s and 1960s, HUG 4212.8, Box 3 (T-Z), folder "WHOI—Suggestions for Director—Confidential," Harvard University Archives.
33. Redfield, *Biographical Memoirs*, 65.
34. Iselin to Bigelow, May 1 and May 21, 1931, AC-9, "Iselin, Columbus (2 of 2), 1930–35," Office of the Director, WHOI Archives.
35. Ibid., May 1.
36. Iselin to Bigelow, July 8, 1931, AC-9, "Iselin, Columbus (2 of 2), 1930–35," Office of the Director, WHOI Archives.
37. Susan Schlee, *On Almost Any Wind: The Saga of the Oceanographic Research Vessel Atlantis* (Ithaca: Cornell University Press, 1978), 32.
38. Ibid., 13.
39. *Annual Announcement: Woods Hole Oceanographic Institution, Second Year 1932–1933*, 4–5.
40. Daniel Behrman, *The New World of the Oceans: Men and Oceanography* (Boston: Little, Brown, 1969), 290.

41. Schlee, *Edge,* 279.

42. *The Woods Hole Oceanographic Institution Report for the Year 1935,* 15.

43. William Wertenbaker, *The Floor of the Sea: Maurice Ewing and the Search to Understand the Earth* (Boston: Little, Brown, 1974), 23.

44. Schlee, *On Almost Any Wind,* 106.

45. Edward C. Bullard, "William Maurice Ewing: May 12, 1906–May 4, 1974," *Biographical Memoirs* 51 (Washington, D.C.: National Academy Press, 1980), 134.

46. Columbus Iselin, "A study of the circulation of the western North Atlantic," *Papers in Physical Oceanography and Meteorology* 4, no. 4 (1936).

47. HUG 4212.7–12, Harvard University Archives.

48. WHOI Oral History Colloquy, July 15, 1991, transcript pp. 23–25.

49. Bigelow to Secretary of the Navy, September 28, 1939, WHOI Archives.

50. Iselin, "A study of the circulation of the western North Atlantic."

51. Vannevar Bush, chairman of the National Advisory Committee on Aeronautics; James B. Conant, president of Harvard University; Richard C. Tolman, dean of CalTech's graduate school; and Karl Compton, president of MIT—all members with Jewett of the National Research Council's Committee on Scientific Aids to Learning. Gary Weir, *An Ocean in Common: American Naval Officers, Scientists, and the Ocean Environment* (Washington, D.C.: Texas A&M University Press, 2001): 97.

52. Iselin to Jewett, January 31, 1941, WHOI Archives.

53. *The Woods Hole Oceanographic Institution Report for the Year 1940,* 9.

54. Laurence Lippsett, ed., *Lamont-Doherty Earth Observatory: Twelve Perspectives on the First Fifty Years, 1949–1999* (Palisades, N.Y.: Lamont-Doherty Earth Observatory, 1999), 23–24.

55. Gordon A. Riley, "Reminiscences of an Oceanographer" (unpublished, about 1984), WHOI Archives, 19.

56. Victoria Kaharl, "Sounding Out the Ocean's Secrets," in series *Beyond Discovery: The Path from Research to Human Benefit* (Washington, D.C.: National Academy of Sciences, 1999), 4.

57. Henry Stommel, "Autobiography," *Collected Works of Henry M. Stommel,* Vol. 1, Nelson G. Hogg and Rui Xin Huang, eds. (Boston: American Meteorological Society, 1995), I-95.

58. This photo is from a manual prepared at WHOI, in 1947, entitled "Instruction Manual for Surface Vessel Bathythermograph and Bathythermograph Winch." AC-06, Box 12, "Historical Instrument Collection," WHOI Archives.

59. Colloquy, July 16, 1991, 13.

60. 1941 correspondence between Iselin and Briscoe and Iselin and Bigelow, WHOI Archives.

61. Bigelow/Iselin Box 6, "Alpha files: R-S (1 of 2) 1941," WHOI Archives.

62. Iselin to Ray D. Wells, WHOI Archives.

63. Riley, "Reminiscences," 45.

64. Columbus O'Donnell Iselin, "Outline of work done at the Woods Hole Oceanographic Institution during the war years" (unpublished, about 1960), WHOI Archives.

65. Iselin, "Personal History," 36–39.

66. Riley, "Reminiscences," 82.

67. John Churchill memorandum to Alfred Redfield, October 13, 1944, WHOI Archives. A first-person account of this experience was written by the ordinary seaman aboard: William Cooper, "*Atlantis* and the Hurricane of 1944," *Woods Hole Reflections,* Mary Lou Smith, ed. (Woods Hole: Woods Hole Historical Collection, 1983, 1997), 235–243.

68. Iselin to Sears, 4 March 1946, Box 8, OD/WHOI [from Weir, *Ocean in Common,* endnote, p. 374].

69. Colloquy, July 16, 1991, afternoon transcript, 42.

70. Iselin, "Personal History," 43–44.

71. Weir, *Ocean in Common,* 230–231.

72. Ibid., 234.

73. Henry M. Stommel, "Columbus O'Donnell Iselin, September 25, 1904–January 5, 1971," *Biographical Memoirs* 64 (Washington, D.C.: National Academy Press, 1994), 175.

74. F. C. (Fritz) Fuglister, "Henry Stommel—On the Light Side," *Evolution of Physical Oceanography: Scientific Surveys in Honor of Henry Stommel,* Bruce Warren and Carl Wunsch, eds., (Cambridge, Mass.: MIT Press, 1981), xxvi–xxvii.

75. Iselin, "Personal History," 14.

76. Riley, "Reminiscences," 81.

77. "Nineteenth Annual Report of the Woods Hole Oceanographic Institution for 1947–48" (unpublished, "Scientific Program," 8), WHOI Archives.

78. Ibid., 17.

79. *Oceanus* (Winter 1952): 1.

80. "Twentieth Annual Report of the Woods Hole Oceanographic Institution for 1948–49" (unpublished, "Floating Facilities," 5), WHOI Archives.

81. Ibid., 1.

82. Ibid., 6.

83. "Twenty-first Annual Report of the Woods Hole Oceanographic Institution for 1949–50" (unpublished, "Develop the Field of Marine Meteorology," 89), WHOI Archives.

84. Iselin, "Personal History," 18.

85. *Woods Hole Currents* 1, no. 2 (1992): 6.

86. Paul Fye, "The Woods Hole Oceanographic Institution: A Commentary," in *Oceanography: The Past,* 3.

87. Edward H. Smith Biographical File, WHOI Archives.

88. Roger Revelle, "The Age of Innocence and War in Oceanography," *Oceans* 1, no. 3 (1969): 15.

89. U.S. Office of Scientific Research and Development, *Science: The Endless Frontier: A Report to the President on a Program for Postwar Scientific Research,* by Vannevar Bush, director, July 1945 (Washington, D.C.: National Science Foundation, 1945).

90. Elizabeth Higgins Gladfelter, *Agassiz's Legacy: Scientists' Reflections on the Value of Field Experience* (New York: Oxford University Press, 2002), 66–67.

91. Papers of Columbus O'D. Iselin, MC-16, "Correspondence (S) 1951–54," WHOI Archives.

92. Columbus O'Donnell Iselin, Notes 1935–1955, WHOI Archives.

93. Kathleen Broome Williams, *Improbable Warriors: Women Scientists and the U.S. Navy in World War II* (Annapolis: Naval Institute Press, 2001), 36.

94. *Oceanus* (Summer 1954): 6.

95. Ibid., 3.

96. J. M. Zeigler, W. D. Athearn, and H. Small, "Profiles Across the Peru-Chile Trench," *Deep-Sea Research* 4 (1957): 238–239.

97. *Oceanus* 2, no. 1 (1953): 20.

98. *Oceanus* (Autumn 1955): 20.

99. Ibid., 5.

100. Paul Fye, "Ocean Policy and Scientific Freedom," Columbus O'Donnell Iselin Memorial Lecture (Marine Technology Society, September 11, 1972, 3). Columbus O'Donnell Iselin Biographical File, WHOI Archives, and *Oceanus* (June 1971): 1.

101. Allyn C. Vine, *Oceanus* (June 1971): 41.

102. Stommel, "Columbus O'Donnell Iselin," 177–178.

103. Raymond Stevens, *Oceanus* (June 1971): 29.

104. Roger Revelle, "The Oceanographic and How It Grew," in *Oceanography: The Past*, 22. George Wood, who was the NAS executive secretary for Project Nobska, wrote in the June 1972 issue of *Oceanus* magazine dedicated to Iselin that "Columbus supplied the [Project Nobska] leadership and stabilizing influence ["oiling the gears," Wood said later in a private communication]; the Associate Director, Ivan Getting—a vice president for research from Raytheon—supplied much of the energy and initiative needed to accomplish what was desired within the time available." Ivan Getting devoted a chapter of his book *All in a Lifetime: Science in the Defense of Democracy* (New York: Vantage Press, 1989) to Project Nobska. One of his footnotes will amuse those familiar with Woods Hole village: "Official visits to Project Nobska by senior flag officers of the navy often supplied a certain amount of levity. When the commander of the Atlantic Fleet came to visit, he arrived in a destroyer that proceeded to anchor in Great Harbor, Woods Hole. In spite of its name, the harbor is called Great only because the adjoining harbor is called Little. In fact, Great Harbor is the mooring area for private yachts. . . . [T]he captain of the destroyer dropped his anchor in the middle of Great Harbor and then watched with horror as his ship was about to sweep out all the yachts in the harbor. Only a swift departure prevented a disaster.

    ". . . As [another admiral's] helicopter approached for a landing on the lawn of the Whitney estate, the downwash of the propeller lifted every dinghy off of every dock in Little Harbor and sent the dinghies down the coast in the currents of Vineyard Sound. . . . The next day or two, a commander visited all the neighbors on Little Harbor, writing out checks to settle damage claims."

105. *Oceanography 1960 to 1970: 1—Introduction and Summary of Recommendations* (Washington, D.C.: National Academy of Sciences–National Research Council, 1959): ii, 0.

106. Peter Tyack, unpublished comments at a memorial for William Watkins, held at WHOI on October 10, 2004.

107. Columbus Iselin, "Short History of the Woods Hole Oceanographic Institution," (unpublished, about 1960), MC-16, "Iselin, C., Lectures and Articles (1 of 3), 1960–65," Office of the Director, WHOI Archives.

108. A December 10, 1954, document authored by Allyn Vine and entitled "Requirements for Directors of Research" reports on a year's worth of staff discussions about "what kind of place [and director] do we want?" It lists seventeen items, each regarding institutional/academic considerations and characteristics for a director. Harvard University Archives 4212.8, "Woods Hole Oceanographic Institution."

109. Jim Luyten and Nelson Hogg, "Foreword," *Oceanus* vol. 35 (Special Issue 1992): 1.

110. The source for the description of this advance in understanding of ocean circulation is largely an article by George Veronis called "Thermal Circulation," *Oceanus* (Summer and Autumn 1957): 19–25.

111. *The Woods Hole Oceanographic Institution Report for the Year 1958*, 12–13.

112. http://www.nas.edu/history/igy/

113. *Oceanus* (December 1960): 18.

114. Selman A. Waksman, *Oceanus* (July 1968—H. B. Bigelow Issue): 10.

115. *Oceanus* (March 1961): 1.

116. *Oceanus* (Winter 1989–88): 103–104.

117. *Oceanus* VI (1; 1958): 7.

118. *Oceanus* (June 1960): 14.

119. Cindy Lee Van Dover tells her story in the delightful book *Deep-Ocean Journeys: Discovering New Life at the Bottom of the Sea* (Reading, Mass.: Addison Wesley, 1996).

120. WHOI Archives folder "Dir. P. Fye, Ship Crawford, 1960–62"

121. and personal communications with Richard Edwards (July 8, 2004) and David Casiles (July 9, 2004).

121. Victoria Kaharl, *Water Baby: The Story of Alvin* (New York: Oxford University Press, 1990).

122. C. Bowin, *Oceanus* (September 1962): 22.

123. Gladfelter, *Agassiz's Legacy*, 111.

124. *1969 Annual Report: Woods Hole Oceanographic Institution*, 50.

125. Memorandum with twenty scientists' signatures to the chairman of the Board of Trustees, November 9, 1961, MC-34, Richard Backus Papers, WHOI Archives.

126. Memorandum from the Executive Committee to the Resident Research Staff, December 19, 1961, MC-34, Richard Backus Papers, WHOI Archives.

127. *Oceanus* (September 1963): 1.

128. Bostwick Ketchum, "The Woods Hole Oceanographic Institution: A Brief History" (unpublished, November 1963, 9–10), WHOI Archives.

129. *1965 Annual Report: Woods Hole Oceanographic Institution*, 25.

130. *Oceanus* XIII, no. 1 (1966): 1.

131. Stommel, "Autobiography," I-51.

132. Luther J. Carter, *Science* 157 (8 September 1967): 1154.

133. *Woods Hole Currents* 8, no. 2 (1999): 6.

134. *Woods Hole Currents* 8, no. 3 (2000): 4.

135. Laurence Lippsett, "Meant for Each Other," *Woods Hole Currents* 8, no. 2 (1999): 7.

136. *1968 Annual Report: Woods Hole Oceanographic Institution*, 15–19.

137. Edward Wenk, Jr., "Genesis of a Marine Policy—the IDOE," *Oceanus* (Spring 1980; "A Decade of Big Science"): 2–11. Wenk was executive secretary to the Marine Council; this article relates the complex background of IDOE and laments its failure to meet "its primary goal of heading off international conflict over intensified use of the sea by using collaborative exploration as a new and untainted means to achieve world comity." The *Oceanus* issue cited provides an interesting retrospective on IDOE.

138. William J. Merrell, Mary Hope Katsouros, and Glen P. Boledovich, "Evolving Institutional Arrangements for U.S. Ocean Sciences," *50 Years of Ocean Discovery: National Science Foundation 1950–2000* (Washington, D.C.: National Academy Press, 2000), 204.

139. Paul M. Fye and William D. Lambert, "Oceanography and the Environment," *Annual Report 1971: Woods Hole Oceanographic Institution*, 17 (for the quotation—the rest of the paragraph is based on the 1970 Director's Report, "Oceanography and Marine Policy").

140. Fye, Iselin Memorial Lecture to Marine Technology Society, 3–7.

141. *1972 Annual Report: Woods Hole Oceanographic Institution*, 42.

142. Fye, Iselin Memorial Lecture to Marine Technology Society, 12.

143. *Woods Hole Notes* 4, no 2 (1972): 4.

144. *Oceanography 1960–1970*.

145. "Paul M. Fye Laboratory of the Woods Hole Oceanographic Institution" (dedication booklet dated June 24, 1983), 3, Paul M. Fye Biographical File, WHOI Archives.

146. The proceedings of these two meetings can be found in *Oceanography: The Past*, edited by Mary Sears and Daniel Merriman, and *Oceanography: The Present and Future*, edited by Peter G. Brewer (New York: Springer-Verlag, 1983).

147. *Woods Hole Currents* 7, no. 2 (1998): 16.

148. "WHOI Director Dorman Resigns," press release, May 18, 1993, Craig E. Dorman Biographical File (1 of 2), WHOI Archives.

Bibliography

Acheson, Roy. *Wickliffe Rose of the Rockefeller Foundation Foundation: 1862–1914, The Formative Years.* Cambridge, Eng.: Killycarn Press, 1992.

Backus, Richard H. Personal communications, 2003–2004.

Behrman, Daniel. *Assault on the Largest Unknown: The International Indian Ocean Expedition 1959–65.* Paris: The Unesco Press, 1981.

———. *The New World of the Oceans: Men and Oceanography.* Boston: Little, Brown, 1969.

Bigelow, Henry B. "A Developing View-Point in Oceanography," *Science,* New Series 71, no. 1830 (Jan. 24, 1930): 84–89.

———. *Memories of a Long and Active Life.* Cambridge, Mass.: The Cosmos Press, 1964.

———. *Oceanography: Its Scope, Problems, and Economic Importance.* Boston: Houghton Mifflin Company, 1931.

Bullard, Edward C. "William Maurice Ewing: May 12, 1906–May 4, 1974," *Biographical Memoirs* 51. Washington, D.C.: National Academy Press, 1980.

Getting, Ivan A. *All in a Lifetime: Science in the Defense of Democracy.* New York: Vantage, 1989.

Gladfelter, Elizabeth Higgins. *Agassiz's Legacy: Scientists' Reflections on the Value of Field Experience.* New York: Oxford University Press, 2002.

Graham, Michael. "Henry Bryant Bigelow." *Deep-Sea Research* 15 (1968).

Hogg, Nelson G., and Rui Xin Huang, eds. *Collected Works of Henry M. Stommel,* Volume I. Boston: American Meteorological Society, 1995.

Innis, Charles S. Personal communications, 2003–2004.

Kaharl, Victoria A. "Sounding Out the Ocean's Secrets." *Beyond Discovery: The Path from Research to Human Benefit.* Washington, D.C.: National Academy of Sciences, 1999.

———. *Water Baby: The Story of Alvin.* New York: Oxford University Press, 1990.

Lillie, Frank R. *The Woods Hole Marine Biological Laboratory.* University of Chicago Press, 1944.

Lippsett, Laurence, ed. *Lamont-Doherty Earth Observatory: Twelve Perspectives on the First Fifty Years, 1949–1999.* Palisades, N.Y.: Lamont-Doherty Earth Observatory, 1999.

Mitchell, James R. Personal communications, July 2004.

Ocean Studies Board, National Research Council. *50 Years of Ocean Discovery: National Science Foundation 1950–2000.* Washington, D.C.: National Academy Press, 2000.

Oceanus Magazine, 1–42 (1952–2000).

Redfield, Alfred C. "Henry Bryant Bigelow, October 3, 1879–December 11, 1967," *Biographical Memoirs* 48. Washington, D.C.: National Academy Press, 1976.

Revelle, Roger. "The Age of Innocence and War in Oceanography," *Oceans* 1:3: 6–16.

———. "Columbus O. Iselin," *Year Book of The American Philosophical Society,* 1977.

Riley, Gordon A. "Reminiscences of an Oceanographer," unpublished, WHOI Archives.

Rockefeller Archive Center.

Schlee, Susan. *The Edge of an Unfamiliar World: A History of Oceanography.* New York: Dutton, 1973.

———. *On Almost Any Wind: The Saga of the Oceanographic Research Vessel* Atlantis. Ithaca: Cornell University Press, 1978.

Scott, David D. Personal communication, June 8, 2004.

Sears, M., and D. Merriman, eds. *Oceanography: The Past.* New York: Springer-Verlag, 1980.

Stommel, Henry M. "Columbus O'Donnell Iselin: September 25, 1904–January 5, 1971." *Biographical Memoirs* 64. Washington, D.C.: National Academy Press, 1994.

Veronis, George. Personal communications, July 2004.

Warren, Bruce A. Personal communications, 2003–2004.

——— and Carl Wunsch, eds. *Evolution of Physical Oceanography: Scientific Surveys in Honor of Henry Stommel.* Cambridge, Mass: MIT Press, 1981.

Weir, Gary E. *An Ocean in Common: American Naval Officers, Scientists, and the Ocean Environment.* College Station: Texas A&M University Press, 2001.

Wertenbaker, William. *The Floor of the Sea: Maurice Ewing and the Search to Understand the Earth.* Boston: Little, Brown, 1974.

Williams, Kathleen Broome. *Improbable Warriors: Women Scientists and the U.S. Navy in World War II.* Annapolis, Md.: Naval Institute Press, 2001.

Wood, George W. Personal communication, July 2004.

Woods Hole Currents 1–10 (1991–2004).

Woods Hole Historical Collection.

Woods Hole Notes 1–18 (1969–1986).

Woods Hole Oceanographic Institution Annual Reports, 1930–2003.

Woods Hole Oceanographic Institution Archives.

Index

Page numbers in italics refer to pictures and information in captions.

GEO. W. ELDRIDGE'S
CHART
J
BUZZARDS BAY
Mishaum Pt.
Hunts Rk
Wilkes Ledge
Sandspit
PENIKESE Id
Gull I.
Middle Ground
Cuttyhunk Hbr.
Cuttyhunk Pond
Pease Ledge
CUTTYHUNK Id
FIX. WHITE
NASHAWENA Id
North Rk
South Rk
PASQUE Id
Quicks Hole
Robinsons Hole
BELL
Canapitsit Channel
Current full strength 5 knots
X.MASS. HUM. SOC.
LIFE SAV. STA.
STATUTE MILES
NAUTICAL MILES
1916 Chart